信息技术人才培养系列规划教材

大 数 据 技 术 实 战 系 列

Hadoop
大数据开发实战

慕课版

学IT有疑问
就找千问千知！

◎ 千锋教育高教产品研发部 编著

人民邮电出版社

北京

图书在版编目（CIP）数据

Hadoop大数据开发实战：慕课版 / 千锋教育高教产品研发部编著. -- 北京：人民邮电出版社，2020.8（2024.6重印）
ISBN 978-7-115-51924-5

Ⅰ. ①H… Ⅱ. ①千… Ⅲ. ①数据处理软件—程序设计 Ⅳ. ①TP274

中国版本图书馆CIP数据核字(2020)第122268号

内 容 提 要

本书全面介绍了Hadoop这一高性能的海量数据处理和分析平台。全书共分11章：第1章首先让读者对大数据及Hadoop有一个总体的认识；第2章主要讲解如何搭建Hadoop集群；第3章～第5章讲解HDFS分布式文件系统、MapReduce分布式计算框架以及ZooKeeper分布式协调服务；第6章讲解Hadoop 2.0的新特性；第7章～第10章主要讲解Hadoop生态圈中的相关辅助系统，包括Hive、HBase分布式存储系统、Flume、Sqoop；第11章讲解了综合项目——电商精准营销，该项目涵盖从前期设计到最终实施的整个过程，对全书知识点进行串联和巩固，使读者加深对Hadoop技术的理解。

本书避免一味地铺陈理论，以实战带动讲解，使读者快速掌握技术并能学以致用。本书可作为普通高等院校的教材及教学参考书，也可作为大数据方向的培训教材，还可作为Hadoop初学者及相关开发人员的参考书。

◆ 编　著　千锋教育高教产品研发部
责任编辑　李　召
责任印制　王　郁　陈　犇

◆ 人民邮电出版社出版发行　北京市丰台区成寿寺路11号
邮编　100164　电子邮件　315@ptpress.com.cn
网址　https://www.ptpress.com.cn
大厂回族自治县聚鑫印刷有限责任公司印刷

◆ 开本：787×1092　1/16
印张：17.5　　　　　　　　2020年8月第1版
字数：476千字　　　　　　　2024年6月河北第9次印刷

定价：59.80元

读者服务热线：(010)81055256　印装质量热线：(010)81055316
反盗版热线：(010)81055315
广告经营许可证：京东市监广登字20170147号

编 委 会

主　编：罗力文　胡耀文　潘庆先

副主编：李亚东　赵　凯　陈峰高　周凤翔　曲北北

　　　　　巩艳华　王　璞　刘　尧　张　涛　杨红艳

编　委：李雪梅　尹少平　张红艳　白妙青　赵　强

　　　　　耿海军　李素清

前言 FOREWORD

当今世界是知识爆炸的世界，科学技术与信息技术快速发展，新型技术层出不穷，教科书也要紧随时代的发展，纳入新知识、新内容。目前很多教科书注重算法讲解，但是，如果在初学者还不会编写一行代码的情况下，教科书就开始讲解算法，会打击初学者学习的积极性，让其难以入门。

党的二十大报告中提到："全面提高人才自主培养质量，着力造就拔尖创新人才，聚天下英才而用之。"IT 行业需要的不是只有理论知识的人才，而是技术过硬、综合能力强的实用型人才。高校毕业生求职面临的第一道门槛就是技能与经验。学校往往注重学生理论知识的学习，忽略了对学生实践能力的培养，导致学生无法将理论知识应用到实际工作中。

为了杜绝这一现象，本书倡导快乐学习、实战就业，在语言描述上力求准确、通俗易懂，在章节编排上循序渐进，在语法阐述中尽量避免术语和公式，从项目开发的实际需求入手，将理论知识与实际应用相结合，目标就是让初学者能够快速成长为初级程序员，积累一定的项目开发经验，从而在职场中拥有一个高起点。

千锋教育

本书特色

在大数据时代，数据的存储与挖掘显得至关重要。企业在追求具备高可靠性、高扩展性及高容错性的大数据处理平台的同时还希望能够降低成本。Hadoop 作为大数据生态系统中的核心框架，专为离线和大规模数据处理而设计，正好解决了企业的实际需求。Hadoop 的核心组件 HDFS 为海量数据提供了分布式存储，MapReduce 则为海量数据提供了分布式计算。很多互联网公司都使用 Hadoop 来实现公司的核心业务，如华为的云计算平台、淘宝的推荐系统等。与海量数据相关的领域都有 Hadoop 的身影。

通过本书你将学习到以下内容。

第 1 章：介绍了大数据的由来及 Hadoop 的特性。

第 2 章：介绍了 Hadoop 集群搭建。

第 3 章～第 5 章：介绍了 HDFS 分布式文件系统、MapReduce 分布式计算框架以及 ZooKeeper 分布式协调服务，带领读者认识 Hadoop 的核心功能。

第 6 章：介绍了 Hadoop 2.0 的新特性。

第 7 章～第 10 章：介绍了 Hadoop 生态圈中的相关辅助系统，包括 Hive、HBase 分布式存储系统、Flume、Sqoop。

第 11 章：讲解了一个综合项目——电商精准营销。读者可以通过该项目体验从前期设计到最终实施的整个过程，使全书知识点融会贯通。

针对高校教师的服务

千锋教育基于多年的教育培训经验，精心设计了"教材+授课资源+考试系统+测试题+辅助案例"教学资源包。教师使用教学资源包可节约备课时间，缓解教学压力，显著提高教学质量。

本书配有千锋教育优秀讲师录制的教学视频，按知识结构体系已部署到教学辅助平台"扣丁学堂"，可以作为教学资源使用，也可以作为备课参考。本书配套教学视频，可登录"扣丁学堂"官方网站下载。

高校教师如需配套教学资源包，也可扫描下方二维码，关注"扣丁学堂"师资服务微信公众号获取。

扣丁学堂

针对高校学生的服务

学 IT 有疑问，就找"千问千知"，这是一个有问必答的 IT 社区，平台上的专业答疑辅导老师承诺在工作时间 3 小时内答复您学习 IT 时遇到的专业问题。读者也可以通过扫描下方的二维码，关注"千问千知"微信公众号，浏览其他学习者在学习中分享的问题和收获。

学习太枯燥，想了解其他学校的伙伴都是怎样学习的？你可以加入"扣丁俱乐部"。"扣丁俱乐部"是千锋教育联合各大校园发起的公益计划，专门面向对 IT 有兴趣的大学生，提供免费的学习资源和问答服务，已有超过 30 万名学习者获益。

千问千知

资源获取方式

本书配套源代码、习题答案的获取方法：读者可添加小千 QQ 号 2133320438 索取，也可登录人邮教育社区 www.ryjiaoyu.com 进行下载。

致谢

本书由千锋教育大数据教学团队整合多年积累的教学实战案例，通过反复修改最终撰写完成。多名院校老师参与了教材的部分编写与指导工作。除此之外，千锋教育的 500 多名学员参与了教材的试读工作，他们站在初学者的角度对教材提出了许多宝贵的修改意见，在此一并表示衷心的感谢。

意见反馈

虽然我们在本书的编写过程中力求完美，但书中难免有不足之处，欢迎读者给予宝贵意见，联系方式：huyaowen@1000phone.com。

千锋教育高教产品研发部

2023 年 5 月于北京

目 录 CONTENTS

第 1 章 初识 Hadoop ············· 1

- 1.1 大数据简介 ················· 1
  - 1.1.1 大数据的五大特征 ············ 1
  - 1.1.2 大数据的六大发展趋势 ········ 3
  - 1.1.3 大数据在电商行业的应用 ······ 4
  - 1.1.4 大数据在交通行业的应用 ······ 5
  - 1.1.5 大数据在医疗行业的应用 ······ 5
- 1.2 大数据技术的核心需求 ············ 5
- 1.3 Hadoop 简介 ··············· 6
  - 1.3.1 什么是 Hadoop ············ 6
  - 1.3.2 Hadoop 的产生和发展 ········ 6
  - 1.3.3 Hadoop 的优缺点 ··········· 7
  - 1.3.4 Hadoop 版本介绍 ··········· 7
  - 1.3.5 Hadoop 生态圈的相关组件 ···· 7
  - 1.3.6 Hadoop 应用介绍 ··········· 8
  - 1.3.7 国内 Hadoop 的就业情况分析 ··· 9
  - 1.3.8 分布式系统概述 ············ 10
- 1.4 离线数据分析流程介绍 ·········· 10
  - 1.4.1 项目需求描述 ············· 11
  - 1.4.2 数据来源 ················· 11
  - 1.4.3 数据处理流程 ············· 11
  - 1.4.4 项目最终效果 ············· 12
- 1.5 大数据学习流程 ··············· 12
- 1.6 本章小结 ···················· 13
- 1.7 习题 ······················· 14

第 2 章 搭建 Hadoop 集群 ····· 15

- 2.1 安装准备 ···················· 15
  - 2.1.1 虚拟机安装 ··············· 15
  - 2.1.2 虚拟机克隆 ··············· 21
  - 2.1.3 Linux 系统网络配置 ········ 23
  - 2.1.4 SSH 服务配置 ············· 26
- 2.2 Linux 基本命令 ··············· 28
  - 2.2.1 系统工作命令 ············· 29
  - 2.2.2 磁盘操作命令 ············· 30
  - 2.2.3 目录与文件操作命令 ········ 30
  - 2.2.4 权限操作命令 ············· 31
- 2.3 Hadoop 集群搭建 ············· 32
  - 2.3.1 Hadoop 集群部署模式 ······· 32
  - 2.3.2 安装 JDK ················· 32
  - 2.3.3 安装 Hadoop ·············· 33
  - 2.3.4 Hadoop 集群配置 ·········· 34
- 2.4 Hadoop 集群测试 ············· 37
  - 2.4.1 格式化文件系统 ············ 37
  - 2.4.2 启动和关闭 Hadoop 进程命令 ··· 37
  - 2.4.3 启动和查看 Hadoop 进程 ····· 38
  - 2.4.4 查看 Web 界面 ············ 38
- 2.5 使用 Hadoop 集群 ············ 39
- 2.6 本章小结 ···················· 40
- 2.7 习题 ······················· 40

第 3 章 HDFS 分布式文件系统 ····················· 41

- 3.1 HDFS 简介 ·················· 41
  - 3.1.1 HDFS 的概念 ·············· 41
  - 3.1.2 HDFS 数据的存储和读取方式 ··· 42
  - 3.1.3 HDFS 的特点 ·············· 42

3.2 HDFS 存储架构和数据读写流程 ····43
　3.2.1 HDFS 的存储架构 ············43
　3.2.2 HDFS 的数据读写流程 ······44
3.3 HDFS 的 Shell 命令 ···············46
3.4 Java 程序操作 HDFS ·············47
　3.4.1 HDFS Java API 概述 ········47
　3.4.2 使用 Java API 操作 HDFS ··47
3.5 Hadoop 序列化 ····················55
　3.5.1 Hadoop 序列化简介 ·········55
　3.5.2 常用实现 Writable 接口的类 ··56
　3.5.3 自定义实现 Writable 接口
　　　 的类 ····························58
3.6 Hadoop 小文件处理 ···············59
　3.6.1 压缩小文件 ····················59
　3.6.2 创建序列文件 ·················60
3.7 通信机制 RPC ·····················63
　3.7.1 RPC 简介 ······················63
　3.7.2 Hadoop 的 RPC 架构 ········63
3.8 本章小结 ···························64
3.9 习题 ·································64

第 4 章 MapReduce 分布式计算框架 ···············65

4.1 认识 MapReduce ··················65
　4.1.1 MapReduce 核心思想 ········65
　4.1.2 MapReduce 编程模型 ········65
　4.1.3 MapReduce 编程案例
　　　 ——WordCount ···············67
4.2 MapReduce 编程组件 ············72
　4.2.1 InputFormat 组件 ·············72
　4.2.2 OutputFormat 组件 ··········73
　4.2.3 RecordReader 组件和
　　　 RecordWriter 组件 ···········76

　4.2.4 Partitioner 组件 ···············76
　4.2.5 Combiner 组件 ················78
4.3 MapReduce 作业解析 ············82
　4.3.1 MapReduce 作业简介 ·······82
　4.3.2 MapReduce 作业运行时的资源
　　　 调度 ·····························82
　4.3.3 MapReduce 作业运行流程 ··83
4.4 MapReduce 工作原理 ············83
　4.4.1 Map 任务工作原理 ··········83
　4.4.2 Reduce 任务工作原理 ·······83
4.5 Shuffle 阶段 ·······················83
　4.5.1 Shuffle 的概念 ················83
　4.5.2 Map 端的 Shuffle ············84
　4.5.3 Reduce 端的 Shuffle ········85
4.6 优化——数据倾斜 ···············85
4.7 MapReduce 典型案例——排序 ··86
　4.7.1 部分排序 ······················86
　4.7.2 全排序 ·························87
4.8 MapReduce 典型案例——倒排
　 索引 ································91
　4.8.1 准备模拟数据 ···············91
　4.8.2 输出数据解析 ···············92
　4.8.3 编写 MapReduce 程序 ······92
4.9 MapReduce 典型案例——连接 ·· 94
　4.9.1 准备模拟数据 ···············94
　4.9.2 输出数据解析 ···············94
　4.9.3 编写 MapReduce 程序 ······94
4.10 MapReduce 典型案例——平均分
　　 以及百分比 ······················97
　4.10.1 准备模拟数据 ···············97
　4.10.2 输出数据解析 ···············97
　4.10.3 编写 MapReduce 程序 ······97

4.11 MapReduce 典型案例——过滤敏感词汇 ·················· 100
    4.11.1 准备模拟数据 ············· 100
    4.11.2 创建敏感词库 ············· 101
    4.11.3 编写 MapReduce 程序 ····· 101
4.12 本章小结 ······················ 103
4.13 习题 ·························· 103

第 5 章 ZooKeeper 分布式协调服务 ················ 105

5.1 认识 ZooKeeper ················ 105
    5.1.1 ZooKeeper 简介 ············ 105
    5.1.2 ZooKeeper 的设计目的 ······ 105
    5.1.3 ZooKeeper 的系统模型 ······ 106
    5.1.4 ZooKeeper 中的角色 ········ 106
    5.1.5 ZooKeeper 的工作原理 ······ 107
5.2 ZooKeeper 安装和常用命令 ······ 108
    5.2.1 ZooKeeper 单机模式 ········ 108
    5.2.2 ZooKeeper 全分布式 ········ 109
    5.2.3 ZooKeeper 服务器常用脚本 ·· 111
    5.2.4 ZooKeeper 客户端节点和命令 ························· 111
5.3 ZooKeeper 客户端编程 ·········· 113
    5.3.1 配置开发环境 ············· 113
    5.3.2 Java 程序操作 ZooKeeper 客户端 ····················· 114
5.4 ZooKeeper 典型应用场景 ········ 115
    5.4.1 数据发布与订阅 ··········· 115
    5.4.2 命名服务 ················· 115
    5.4.3 分布式锁 ················· 116
5.5 本章小结 ······················ 116
5.6 习题 ·························· 116

第 6 章 Hadoop 2.0 新特性 ·············· 118

6.1 Hadoop 2.0 的改进 ·············· 118
    6.1.1 HDFS 存在的问题 ········· 118
    6.1.2 MapReduce 存在的问题 ····· 118
    6.1.3 HDFS 2.0 解决 HDFS 1.0 中的问题 ······················· 119
6.2 YARN 资源管理框架 ············ 119
    6.2.1 YARN 简介 ··············· 119
    6.2.2 YARN 架构 ··············· 119
    6.2.3 YARN 的优势 ············· 120
6.3 Hadoop 的 HA 模式 ············ 120
    6.3.1 HA 模式简介 ············· 120
    6.3.2 HDFS 的 HA 模式 ········· 121
    6.3.3 YARN 的 HA 模式 ········· 127
    6.3.4 启动和关闭 Hadoop 的 HA 模式 ····················· 131
6.4 本章小结 ······················ 132
6.5 习题 ·························· 132

第 7 章 Hive ·············· 133

7.1 数据仓库简介 ·················· 133
    7.1.1 数据仓库概述 ············· 133
    7.1.2 数据仓库的使用 ··········· 133
    7.1.3 数据仓库的特点 ··········· 134
    7.1.4 主流的数据仓库 ··········· 134
7.2 认识 Hive ······················ 134
    7.2.1 Hive 简介 ················ 134
    7.2.2 Hive 架构 ················ 135
    7.2.3 Hive 和关系型数据库比较 ·· 136
7.3 Hive 安装 ······················ 136

7.4 Hive 数据类型	140
7.4.1 Hive 基本数据类型	140
7.4.2 Hive 复杂数据类型	141
7.5 Hive 数据库操作	142
7.6 Hive 表	143
7.6.1 内部表和外部表	143
7.6.2 对表进行分区	149
7.6.3 对表或分区进行桶操作	153
7.7 Hive 表的查询	156
7.7.1 select 查询语句	156
7.7.2 视图	161
7.7.3 Join	162
7.8 Hive 函数	165
7.8.1 Hive 内置函数	165
7.8.2 通过 JDBC 驱动程序使用 Hiveserver2 服务	167
7.8.3 Hive 用户自定义函数	169
7.9 Hive 性能优化	171
7.10 Hive 案例分析	173
7.11 本章小结	174
7.12 习题	174

第8章 HBase 分布式存储系统 ············ 175

8.1 认识 HBase	175
8.1.1 HBase 简介	175
8.1.2 HBase 的数据模型	176
8.1.3 HBase 架构	176
8.1.4 HBase 文件存储格式	178
8.1.5 HBase 存储流程	179
8.1.6 HBase 和 HDFS	179
8.2 HBase 表设计	179

8.2.1 列簇设计	179
8.2.2 行键设计	180
8.3 HBase 安装	180
8.3.1 HBase 的单机模式	180
8.3.2 HBase 的 HA 模式	182
8.4 HBase Shell 常用操作	184
8.5 HBase 编程	190
8.5.1 配置开发环境	190
8.5.2 使用 Java API 操作 HBase	191
8.5.3 使用 HBase 实现 WordCount	193
8.6 HBase 过滤器和比较器	195
8.6.1 过滤器	195
8.6.2 比较器	196
8.6.3 编程实例	196
8.7 HBase 与 Hive 结合	201
8.7.1 HBase 与 Hive 结合的原因	201
8.7.2 Hive 关联 HBase	201
8.8 HBase 性能优化	202
8.9 本章小结	204
8.10 习题	204

第9章 Flume ············ 205

9.1 认识 Flume	205
9.1.1 Flume 简介	205
9.1.2 Flume 的特点	205
9.2 Flume 基本组件	206
9.2.1 Event	206
9.2.2 Agent	206
9.3 Flume 安装	207
9.4 Flume 数据流模型	208
9.5 Flume 的可靠性保证	210
9.5.1 负载均衡	210

9.5.2 故障转移 ………………… 211
9.6 Flume 拦截器 ………………… 212
9.7 采集案例 ……………………… 214
　9.7.1 采集目录到 HDFS ……… 214
　9.7.2 采集文件到 HDFS ……… 215
9.8 本章小结 ……………………… 216
9.9 习题 …………………………… 216

第 10 章 Sqoop …………… 217

10.1 认识 Sqoop ………………… 217
　10.1.1 Sqoop 简介 …………… 217
　10.1.2 Sqoop 原理 …………… 218
　10.1.3 Sqoop 架构 …………… 218
10.2 Sqoop 安装 ………………… 218
10.3 Sqoop 命令 ………………… 220
　10.3.1 Sqoop 数据库连接参数 …… 221
　10.3.2 Sqoop export 参数 …… 221
　10.3.3 Sqoop import 参数 …… 221
　10.3.4 Sqoop import 命令的
　　　　 基本操作 ……………… 221
10.4 Sqoop 数据导入 …………… 222
　10.4.1 将 MySQL 的数据导入
　　　　 HDFS …………………… 222
　10.4.2 将 MySQL 的数据导入
　　　　 Hive ……………………… 223
　10.4.3 将 MySQL 的数据导入
　　　　 HBase …………………… 226
　10.4.4 增量导入 ……………… 227
　10.4.5 按需导入 ……………… 229
10.5 Sqoop 数据导出 …………… 230
　10.5.1 将 HDFS 的数据导出到
　　　　 MySQL ………………… 230

　10.5.2 将 Hive 的数据导出到
　　　　 MySQL ………………… 231
　10.5.3 将 HBase 的数据导出到
　　　　 MySQL ………………… 231
10.6 Sqoop job …………………… 233
10.7 本章小结 …………………… 233
10.8 习题 ………………………… 234

第 11 章 综合项目——电商精准营销 ………… 235

11.1 项目概述 …………………… 235
　11.1.1 项目背景介绍 ………… 235
　11.1.2 项目架构设计 ………… 235
11.2 项目详细介绍 ……………… 237
　11.2.1 项目核心关注点 ……… 237
　11.2.2 重要概念 ……………… 237
　11.2.3 维度 …………………… 238
11.3 项目模块分析 ……………… 239
　11.3.1 用户基本信息分析模块 … 239
　11.3.2 浏览器分析模块 ……… 239
　11.3.3 地域分析模块 ………… 239
　11.3.4 外链分析模块 ………… 239
11.4 数据采集 …………………… 240
　11.4.1 日志采集系统概述 …… 240
　11.4.2 JS SDK 收集数据 …… 240
　11.4.3 Java SDK 收集数据 … 242
　11.4.4 使用 Flume 搭建日志采集
　　　　 系统 …………………… 243
　11.4.5 日志信息说明 ………… 244
11.5 数据清洗 …………………… 245
　11.5.1 分析需要清洗的数据 … 245
　11.5.2 解析数据格式转换 …… 245

11.5.3 利用 MapReduce 清洗数据 …… 245
11.6 使用数据仓库进行数据分析 …… 253
　11.6.1 事件板块数据分析 …… 253
　11.6.2 订单板块数据分析 …… 257
　11.6.3 时间板块数据分析 …… 262
11.7 可视化 …… 264
　11.7.1 ECharts 简介 …… 264
　11.7.2 ECharts 的优点 …… 265
　11.7.3 操作流程 …… 265
11.8 本章小结 …… 267
11.9 习题 …… 267

附录 …… 268

第1章 初识Hadoop

本章学习目标
- 了解大数据的概念
- 熟悉大数据的应用场景
- 了解 Hadoop 框架
- 了解大数据的学习流程

随着新一代信息技术的迅猛发展和深入应用，数据的规模不断扩大，数据日益成为继土地、资本之后的又一种重要生产要素，是各个国家和地区争夺的重要资源，谁掌握数据方面的主动权和主导权，谁就能赢得未来。美国奥巴马政府将数据定义为"未来的新石油"，认为一个国家拥有数据的规模、活性及解释运用数据的能力将成为综合国力的重要组成部分，对数据的占有和控制将成为陆权、海权、空权之外的另一个国家核心权力。一个全新的概念——大数据开始风靡全球。

1.1 大数据简介

从前，人们用饲养的马来拉货物。当一匹马拉不动一车货物时，人们不曾想过培育一匹更大更壮的马，而是利用更多的马。同样，当一台计算机无法进行海量数据计算时，人们也无须去开发一台超级计算机，而应尝试着使用更多计算机。

下面来看一组令人瞠目结舌的数据：2018 年 11 月 11 日，支付宝总交易额达到 2135 亿元，支付宝实时计算处理峰值为 17.18 亿条/秒，天猫物流订单量超过 10 亿……

这场狂欢的背后是金融科技的护航，正是因为阿里巴巴公司拥有中国首个具有自主知识产权、全球首个应用在金融核心业务的分布式数据库平台 OceanBase，海量交易才得以有序地进行。分布式集群具有高性能、高并发、高一致性、高可用性等优势，远远超出单台计算机的能力范畴。

1.1.1 大数据的五大特征

大数据（Big Data），是指数据量巨大，无法使用传统工具进行处理的数据集合。通常认为，大数据的典型特征主要体现在以下 5 个方面：大量（Volume）、高速性（Velocity）、多样性（Variety）、价值（Value）、真实性（Veracity），即所谓的"5V"，如图 1.1 所示。

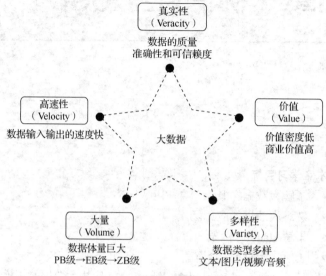

图 1.1 大数据"5V"特征

1. 大量（Volume）

大数据的特征首先就是数据规模巨大。随着互联网、物联网、移动互联等技术的发展，人和事物的所有轨迹都可以被记录下来，数据呈现爆发性增长，需要分析处理的数据量达到 PB、EB，乃至 ZB 级。数据相关计量单位的换算关系如表 1.1 所示。

表 1.1　　　　　　　　　　　　　单位换算关系

单位	换算公式
Byte	1 Byte = 8 bit
KB	1 KB = 1024 Byte
MB	1 MB = 1024 KB
GB	1 GB = 1024 MB
TB	1 TB = 1024 GB
PB	1 PB = 1024 TB
EB	1 EB = 1024 PB
ZB	1 ZB = 1024 EB

2. 高速性（Velocity）

数据的增长速度和处理速度是大数据高速性的重要体现。生活中每个人都离不开互联网，也就是说每个人每天都在向大数据提供大量的资料。与以往的报纸、书信等传统数据载体的传播方式不同，在大数据时代，数据的交换和传播主要通过互联网和云计算等方式实现，速度惊人。正因为如此，大数据对处理和响应速度要求极高，一条数据的分析必须在几秒内完成，数据处理与丢弃几乎无延迟。

3. 多样性（Variety）

大数据具有多样性，不同的数据源产生海量的非结构化数据。大数据可以分为三类，一是结构化数据，如财务系统数据、信息管理系统数据、医疗系统数据等，其特点是数据间因果关系强；二是半结构化数据,如 HTML 文档、邮件、网页等,其特点是数据间因果关系弱;三是非结构化数据,如视频、图片、音频、文本等,其特点是数据间没有因果关系。半结构化、非结构化数据需要经过清洗、整理、筛选,变为结构化数据。

4. **价值**(Value)

大数据的核心特征是价值密度低。由于数据样本不全面,数据采集不及时,数据不连续等原因,有价值的数据所占比例很小。与传统的小数据相比,大数据最大的价值在于,可以从大量不相关的各种类型的数据中,挖掘出对未来趋势与模式预测分析有用的信息,通过机器学习、人工智能或数据挖掘等方法深度分析,得到新规律和新知识,并运用于交通、电商、医疗等各个领域,最终达到提高生产效率、推进科学研究的效果。

5. **真实性**(Veracity)

大数据的重要性就在于对决策的支持,数据的规模并不能决定其能否为决策提供帮助,数据的真实性和质量才是成功决策最坚实的基础。真实是对大数据的重要要求,也是大数据面临的巨大挑战。

1.1.2 大数据的六大发展趋势

虽然现在大数据仍处在起步阶段,还面临着很多的困难与挑战,但大数据发展前景是非常可观的。

1. **数据将呈现指数级增长**

近年来,随着移动互联、社交网络、电子商务和云计算的兴起,日志、图片、音频、视频等各类数据呈指数级增长。有关资料显示,2011 年,全球数据规模为 1.8ZB,可以填满 575 亿个 32GB 的 iPad。到 2020 年,全球数据将达到 40ZB,如果把它们全部存入蓝光光盘,这些光盘和 424 艘尼米兹级航母重量相当。美国互联网数据中心则指出,互联网上的数据每年增长 50%,目前世界上 90%以上的数据是最近几年产生的。

2. **数据将成为最有价值的资源**

在大数据时代,数据成为继土地、劳动、资本之后的新要素,数据与它们共同构成企业未来发展的核心竞争力。《华尔街日报》在一份题为"大数据,大影响"的报告中提到,数据已经成为一种新的资产类别,就像货币或黄金一样。IBM 执行总裁罗睿兰指出:"数据将成为一切行业当中决定胜负的根本因素,并最终成为人类至关重要的自然资源。"随着大数据应用的不断发展,我们有理由相信大数据将成为机构和企业的重要资产和争夺的焦点,谷歌、苹果、亚马逊、阿里巴巴、腾讯等互联网公司正在运用大数据力量获得商业上更大的成功,并且将会继续通过大数据来提升自己的竞争力。

3. **大数据和传统行业智能融合**

通过对大数据的收集、整理、分析、挖掘,不仅可以发现城市治理症结,掌握经济运行趋势,还能够驱动精确设计和精确生产模式,引领服务业的精确化和增值化,创造互动的创意产业新形态。麦当劳、肯德基以及苹果等专卖店的位置都是建立在数据分析基础之上的精准选址。百度、阿里巴巴、腾讯等通过对海量数据的掌握和分析,为用户提供更加专业化和个性化的服务。在智慧城市建设不断深入的情况下,大数据必将在智慧城市中发挥越来越重要的作用。由城市数字化到智慧城市,关键是要实现对数字信息的智慧处理,其核心就是大数据处理技术,大数据将成为智慧城市的核心引擎。智慧金融、智慧安防、智慧医疗、智慧教育、智慧交通、智慧城管等,无不是大数据和传统产业融合的重要领域。

4. 数据将越来越开放

大数据是人类的共同资源、共同财富，数据开放共享是不可逆转的历史潮流。随着各国政府和企业对开放数据的社会效益和商业价值认识的不断提升，全球必将掀起数据开放的热潮。事实上，大数据的发展需要全世界、全人类共同协作，变私有大数据为公共大数据，最终实现私有、企业自有、行业自有大数据的全球性整合，才不致形成一个个毫无价值的"数据孤岛"。大数据越关联越有价值，越开放越有价值。尤其是公共事业和互联网企业的开放数据将越来越多。目前，一些发达国家已经在政府和公共事业的数据开放方面做出了表率。中国政府也将一方面带头力促数据共享，另一方面通过推动建设各类大数据服务交易平台，为数据使用者提供丰富的数据来源和应用。

5. 大数据安全将备受重视

大数据在经济社会中得到广泛应用的同时，其安全性也必将受到更多的重视。大数据时代，在我们用数据分析和数据挖掘等大数据技术获取有价值信息的同时，"黑客"也可以利用这些大数据技术最大限度地收集有用的信息，对其感兴趣的目标发起更加精准的攻击。近年来，个人隐私、企业商业信息甚至国家机密的泄露事件时有发生，一些国家纷纷制定完善了保护信息安全、防止隐私泄露的法律法规。可以预见，在不久的将来，其他国家也会迅速跟进，以更好地保障本国政府、企业乃至居民的数据安全。

6. 大数据人才将备受欢迎

随着大数据技术的不断发展，大数据应用的日益广泛，大数据分析师、大数据算法工程师、数据管理专家、数据产品经理等经验丰富的数据分析人员将成为社会的稀缺资源和各机构争夺的人才。2018年中国大数据领域的复合型人才缺口约为160万。从2016年开始，国家为应对大数据人才市场缺口，创设了"数据科学与大数据技术"本科专业。截至2018年，通过审批设置该专业的学校数量从2016年的3所增长到283所，增长了近百倍。

1.1.3 大数据在电商行业的应用

当用户在一些电商平台购买商品时，将商品加入购物车，会显示购买了该商品的用户还购买了哪些商品，如图1.2所示。结算时，又会显示"猜你喜欢"的商品信息，如图1.3所示。这些功能都用到了大数据技术，是对大量的用户浏览记录进行统计分析后做出的精准推送。

图1.2 购买商品

图 1.3　猜你喜欢

精准广告的推送依赖于对海量互联网用户的相关数据的统计分析，其核心是用户画像，这就需要一个分布式的快速响应的数据库系统。

1.1.4　大数据在交通行业的应用

目前，出行难的问题在各大城市都亟待解决，好在现在可以利用先进的传感技术、网络技术、计算技术、控制技术、智能技术，对道路交通进行全面的监控和疏导。在大数据时代，智慧交通需要融合传感器、监控视频和 GPS 等设备产生的海量数据，甚至参考气象监测设备产生的天气数据，从中提取出人们真正需要的信息，及时而准确地进行发布和传送，并通过计算直接提供最佳出行方式和路线。

1.1.5　大数据在医疗行业的应用

除了新兴行业，传统行业也需要大数据，例如，医疗行业要分析大量的病例、买药行为记录、诊断数据。在大数据时代，可以将医疗机构的电子病历标准化，形成全方位、多维度的大数据仓库。系统首先全面分析患者的基本资料、诊断结果、药方、医疗保险情况和付款记录等诸多数据，然后将分析结果综合起来，在医生的参与下通过决策支持系统选择最佳医疗解决方案。

1.2　大数据技术的核心需求

大数据技术需要解决两个核心需求：一个是数据存储；另一个是数据运算。

数据存储：将数据（文件）分散到一个集群的多台机器上存储。

数据运算：数据分析要通过程序来实现，程序的运行最终又是机器的 CPU、内存、磁盘等硬件的运行，这些运行就是运算。

简单点理解，大数据技术就是将大量的数据分割成多个小块，由多台计算机分工计算，最后将结果汇总。执行分布式计算的计算机总称集群，如果把人和计算机做类比，那么集群就是一个团队，如图 1.4 所示。单兵作战的时代已经过去，团队合作才是王道。

为什么需要分布式计算？因为"大数据"来了，单个计算机不够用了，即数据量远远超出单个计算机的处理能力范围：有时是单位时间内的数据量大，比如12306网站每秒可能有数以万计的访问；有时是数据总量大，比如百度搜索引擎要在服务器上检索数亿条中文网页信息。

在大数据体系下，一切数据运算逻辑的实现，都要依靠分布式计算系统。实现分布式计算的方案有很多，真正让大数据技术大踏步前进的是按照Google理论实现的开源免费产品Hadoop，目前已经形成了以Hadoop为核心的大数据技术生态圈。

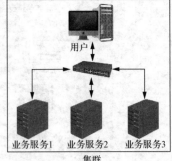

图1.4 分布式计算

1.3 Hadoop 简介

1.3.1 什么是Hadoop

Hadoop是Apache旗下的开源软件平台，是一种分布式框架，可利用服务器集群，根据用户的自定义业务逻辑，对海量数据进行分布式处理。

Hadoop有三个基本组件：HDFS（Hadoop Distributed File System，Hadoop分布式文件系统）负责分布式存储，MapReduce负责分布式计算，YARN（Yet Another Resource Negotiator，另一种资源协调者）为MapReduce提供硬件资源调度。但是Hadoop的核心组件处于底层，直接基于这个底层框架来设计数据分析逻辑结构比较烦琐，开发效率很低，所以在Hadoop框架之上，又衍生了Hive这样的快捷开发工具，后面的章节会详细讲解。

广义上说，Hadoop通常是一个更广泛的概念——Hadoop生态圈，包括Hadoop、Hive、HBase、Flume、Kafka、Sqoop、Spark、Flink等。

1.3.2 Hadoop的产生和发展

Hadoop最早起源于Nutch。Nutch是一个Java实现的开源搜索引擎。Nutch的设计目标是构建一个大型的全网搜索引擎，这个搜索引擎具备网页抓取、索引、查询等功能。但随着抓取网页数量的增加，Nutch遇到了严重的问题——如何存储数十亿个网页和对信息建立索引。

2003年、2004年谷歌发表的两篇论文为以上问题提供了解决方案。论文内容主要涉及以下两个框架。

（1）分布式文件系统（Google File System，GFS），主要用于海量数据的存储。

（2）分布式计算框架（MapReduce），主要用于海量数据的索引计算。

Nutch的开发人员根据谷歌的GFS和MapReduce，完成了开源版本的NDFS（Nutch Distributed File System）和MapReduce。2006年2月，Nutch的NDFS和MapReduce发展成独立的项目Hadoop。

2008年，Hadoop成为Apache的顶级项目。同年，Hadoop成为速度最快的TB级数据排序系统。自此以后，Hadoop逐渐被企业应用于生产，处理大数据的速度越来越快。目前，Hadoop已经被主流企业广泛使用。

1.3.3 Hadoop 的优缺点

1. Hadoop 的优点

（1）Hadoop 具有按位存储和数据处理的高可靠性。

（2）Hadoop 通过可用的计算机集群分配数据，完成存储和计算任务，这些集群可以扩展至数以千计的节点，具有高扩展性。

（3）Hadoop 能够在节点之间移动数据，并保证各个节点的动态平衡，处理速度非常快，具有高效性。

（4）Hadoop 能够自动保存数据的多个副本，并自动将失败的任务重新分配，具有高容错性。

2. Hadoop 的缺点

（1）Hadoop 不适用于低延迟数据访问。

（2）Hadoop 不能高效存储大量小文件。

（3）Hadoop 不支持多用户写入并任意修改文件。

1.3.4 Hadoop 版本介绍

Hadoop 自诞生以来，主要出现了 Hadoop 1.0、Hadoop 2.0、Hadoop 3.0 三个系列多个版本。

HDFS 和 MapReduce 是 Hadoop 1.0 的核心组件，Hadoop 生态圈里的很多组件都是基于 HDFS 和 MapReduce 发展而来的。Hadoop 1.0 之后出现了 Hadoop 2.0，Hadoop 2.0 在 Hadoop 1.0 的基础上做了改进。Hadoop 2.0 的三大核心组件分别是 HDFS、MapReduce、YARN。目前绝大部分企业使用的是 Hadoop 2.0，本书采用的是 Hadoop 2.7.3 这一版本。

Hadoop 2.0 的一个公共模块和三大核心组件组成了四个模块，简介如下。

（1）Hadoop Common：为其他 Hadoop 模块提供基础设施。

（2）HDFS：具有高可靠性、高吞吐量的分布式文件系统。

（3）MapReduce：基于 YARN 系统的分布式离线并行计算框架。

（4）YARN：负责作业调度与集群资源管理的框架。

1.3.5 Hadoop 生态圈的相关组件

除了 HDFS、MapReduce、YARN 三大核心组件外，Hadoop 生态圈的其他组件还有 ZooKeeper、MySQL、Hive、HBase、Flume、Sqoop，如图 1.5 所示。Hadoop 生态圈各组件的说明，如表 1.2 所示。

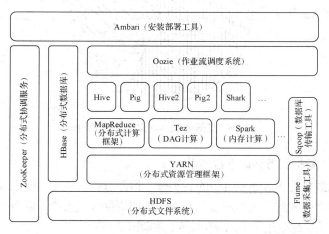

图 1.5 Hadoop 生态圈

表1.2　　Hadoop生态圈各组件的说明

组件	说明
HDFS	分布式文件系统
MapReduce	分布式计算框架
YARN	分布式资源管理框架
ZooKeeper	分布式协调服务
Oozie	作业流调度系统
Hive	数据仓库工具
HBase	分布式数据库
Flume	数据采集工具
Sqoop	数据传输工具

这些组件在后面的章节中会详细讲解。

1.3.6　Hadoop应用介绍

1. Hadoop用于用户画像

用户画像是真实用户的虚拟代表，是建立在一系列真实数据之上的目标用户模型。通过对客户多方面的了解，将多种信息集合在一起，总结出一定的特征与气质，就形成了用户的独特"画像"。

2. Hadoop用于数据挖掘

金融行业的个人征信分析，证券行业的投资模型分析，交通行业的车辆、路况监控分析，电信行业的用户上网行为分析……Hadoop并不会跟某种具体的行业或者某个具体的业务挂钩，它只是一种用来做数据挖掘的工具。某网站点击流日志数据挖掘饼图如图1.6所示。

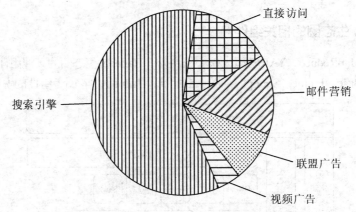

图1.6　某网站点击流日志数据挖掘饼图

3. Hadoop应用于数据服务基础平台建设

一个成熟的大数据开发平台必不可少的各类核心组件有：工作流调度系统、集成开发环境、元数据管理系统、数据交换服务、数据可视化服务、数据质量管理服务，以及测试环境的建设等。数据服务基础平台的建设广泛运用了Hadoop技术，如图1.7所示。

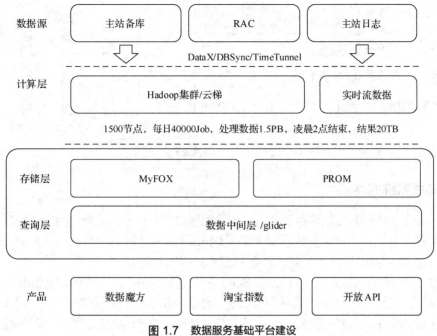

图 1.7 数据服务基础平台建设

1.3.7 国内 Hadoop 的就业情况分析

1. Hadoop 就业整体情况

（1）大数据产业已纳入国家"十三五"规划。
（2）各大城市都在进行智慧城市项目建设，而智慧城市的根基就是大数据综合平台。
（3）互联网时代数据的种类、数量都呈现爆发式增长，各行业对数据的价值日益重视。
（4）相对于传统 Java EE 技术领域，大数据技术领域人才稀缺。
（5）随着现代社会的发展，数据处理和数据挖掘的重要性只会增不会减，因此，大数据技术是一个尚在蓬勃发展且具有广阔前景的领域。

2. Hadoop 就业职位要求

大数据技术是个复合专业，包含应用开发、软件平台、算法、数据挖掘等。因此，大数据技术领域的就业选择是多样的。就 Hadoop 而言，技术人员通常需要具备以下技能或知识。

（1）Hadoop 分布式集群的平台搭建。
（2）Hadoop 分布式文件系统（HDFS）的原理理解及使用。
（3）Hadoop 分布式计算框架（MapReduce）的原理理解及编程。
（4）Hive 数据仓库工具的熟练应用。
（5）Flume、Sqoop、Oozie 等辅助工具的熟练使用。
（6）Shell、Python 等脚本语言的熟练使用。

3. Hadoop 相关职位的薪资水平

Hadoop 的人才需求目前主要集中在北上广深等一线城市，从业者薪资待遇普遍高于传统 Java EE 开发人员，如图 1.8 所示。

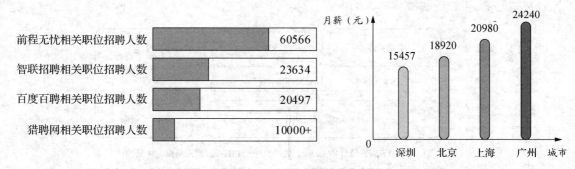

图 1.8　Hadoop 就业情况

1.3.8　分布式系统概述

分布式系统是一个硬件或软件组件分布在不同的网络计算机上，彼此之间仅仅通过消息传递进行通信和协调的系统。由于大数据技术领域的各类技术框架基本上都是分布式系统的，因此，要理解 Hadoop、Storm、Spark 等技术框架，需要先理解基本的分布式系统。

分布式系统，简单来说就是一群独立计算机共同对外提供服务，但是对于系统的用户来说，就像一台计算机在提供服务一样。分布式意味着可以采用更多的普通计算机（相对于昂贵的大型机）组成分布式集群对外提供服务，计算机越多，CPU、存储资源等也就越多，能够处理的并发访问量也越大。

在分布式系统中，计算机之间的通信和协调主要通过网络进行，所以在空间上几乎没有任何限制，这些计算机可能被放在不同的机柜上，也可能被部署在不同的机房中，还可能在不同的城市里，甚至分布在不同的国家和地区。

1.4　离线数据分析流程介绍

本节我们通过综合项目"网站或 App 点击流日志数据挖掘系统"感受数据分析系统的宏观概念及处理流程，初步理解 Hadoop 等框架在其中的应用环节。

应用广泛的数据分析系统"Web 日志数据挖掘"如图 1.9 所示。

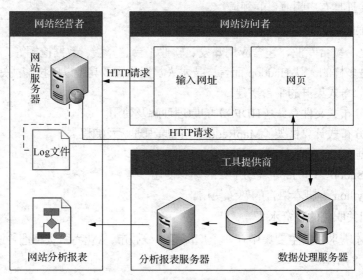

图 1.9　Web 日志数据挖掘

1.4.1 项目需求描述

点击流日志包含着网站运营的重要信息，通过日志分析，可以知道网站的访问量、哪个网页访问人数最多、哪个网页最有价值，了解广告转化率、访客的来源信息、访客的终端信息等。

1.4.2 数据来源

本项目的数据主要来自用户的点击行为。获取方式：在页面预埋一段 Java Script 程序，为页面上想要监听的标签绑定事件，只要用户点击或移动到标签，即发送 Ajax 请求到后台 Servlet 程序，用 Log4j 记录下事件信息，从而在 Web 服务器（Nginx、Tomcat 等）上形成不断增长的日志文件。

1.4.3 数据处理流程

1. 流程图解析

本项目与典型的 BI（Business Intelligence，商业智能）系统极其相似，整体流程如图 1.10 所示。

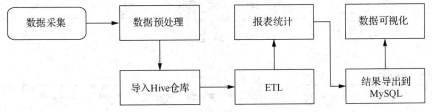

图 1.10 整体流程图

但是，由于本项目的前提是处理海量数据，因此，流程中各环节所使用的技术跟传统 BI 系统完全不同，这些技术后面会一一讲解。

- 数据采集：定制开发采集程序，或使用开源框架 Flume。
- 数据预处理：定制开发 MapReduce 程序运行于 Hadoop 集群。
- 数据仓库技术：基于 Hadoop 的 Hive。
- 数据导出：基于 Hadoop 的 Sqoop 数据导入导出工具。
- 数据可视化：定制开发 Web 程序或使用 Kettle 等产品。
- 整个过程的流程调度：Hadoop 生态圈中的 Oozie 工具或其他类似开源产品。

2. 项目技术架构图

项目技术架构图如图 1.11 所示。

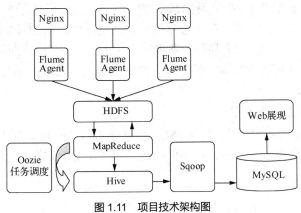

图 1.11 项目技术架构图

1.4.4 项目最终效果

经过完整的数据处理流程后，系统会周期性输出各类统计指标的报表，在生产实践中，最终需要将这些报表数据以可视化的形式展现出来，本项目采用 Web 程序来实现数据可视化，效果如图 1.12 所示。

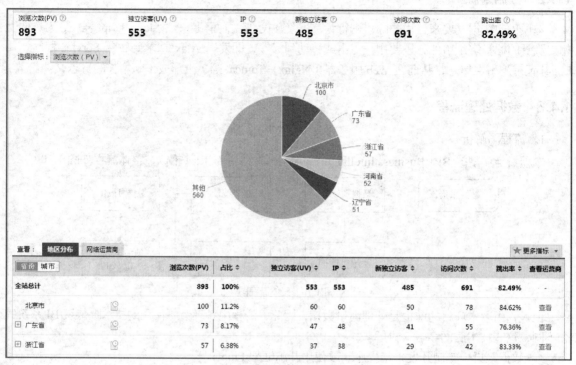

图 1.12 项目最终效果

1.5 大数据学习流程

大数据技术是一门综合技术，要求开发者既具备良好的 Java 基础，又对数据敏感，并掌握主流的大数据开发、数据挖掘和机器学习等技能。在 Java 基础学习阶段，学完 Java 的基本语法后，应尝试用 Java 完成分析案例，目的是对前面所学的知识进行巩固，活学活用，提高编程能力和对数据的敏感度。在大数据技术学习阶段，目前最流行的莫过于学习 Hadoop、Spark 和 ElasticSearch 等大型技术栈，对于这些大数据技术，既要熟练掌握原理和开发、部署方法，更要通过大量真实的大数据实战项目，来加强业务理解和实战开发能力，从而快速领会项目实战技巧，快速积累经验，拥有胜任高端大数据开发工程师岗位的能力。大数据开发工程师需要具备的技能如图 1.13 所示。

1. 第一阶段：大数据准备工作

学习大数据首先要有 Java 语言基础、MySQL 数据库基础并对 Java Web 有一定的了解，没有学习过的读者可参考千锋系列图书《Java 语言程序设计》《MySQL 数据库从入门到精通》《Java Web 开发实战》。

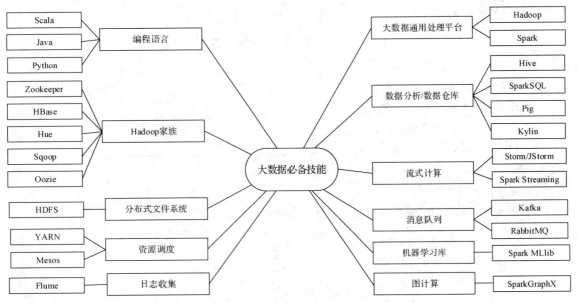

图 1.13　大数据开发工程师技能图

2. 第二阶段：学习大数据必备技能

第二阶段需要学习本书重点讲解的 Hadoop。本书首先讲解了 Hadoop 集群搭建和基础知识，接着对 Hadoop 生态圈中的重要组件进行逐一介绍，并以实战案例贯穿讲解。学完本书，读者可以深刻理解大数据思想。

千锋大数据系列图书还有另外两本，其中一本是《Spark 大数据深入解析》。Hadoop 是基于磁盘的，它的运算结果保存在磁盘当中；而 Spark 的运算是基于内存的。因此 Spark 的运算速度是 Hadoop 的 100 倍。

3. 第三阶段：项目提升

Hadoop 具备 Spark 所没有的功能特性，如分布式文件系统等。而 Spark 为需要它的那些数据集提供了实时内存处理。完美的大数据场景正是设计人员当初预想的那样：让 Hadoop 和 Spark 在同一个团队里面协同运行。

千锋大数据系列图书的另外一本是《大数据企业案例综合实战》，通过使用 Hadoop 和 Spark 开发大数据企业实战项目，贯穿运用前面所学到的知识，进行反复训练，培养独立解决问题的能力。

4. 第四阶段：后续提高

真正的数据科学家能够将大数据与人工智能、机器学习、算法相结合，打通了数据科学的任督二脉，便可以达到技术专家级别。

1.6　本章小结

本章主要对什么是大数据、大数据在日常生活中的应用场景，以及 Hadoop 的背景进行了讲解，并用一个案例进行更详细的说明。

1.7 习题

1. 填空题

（1）大数据技术的两个核心需求：_____、_____。
（2）大数据的"5V"特征包含_____、_____、_____、_____、_____。
（3）Hadoop 1.0 的核心组件为_____、_____。
（4）Hadoop 2.0 的三大核心组件为_____、_____、_____。
（5）目前 Apache Hadoop 发布的版本主要有_____、_____、_____。

2. 思考题

（1）简述什么是大数据。
（2）简述大数据学习流程，以及需要掌握的知识。

第2章 搭建Hadoop集群

本章学习目标
- 了解虚拟机的安装和克隆
- 掌握 Linux 基本命令的用法
- 熟悉 Linux 系统网络配置
- 掌握 Hadoop 集群的搭建和配置方法
- 熟悉 Hadoop 集群案例操作

"工欲善其事，必先利其器"，在深入学习 Hadoop、掌握其相关应用前，需要学会搭建集群环境。本章将从虚拟机的安装和配置、Linux 基本命令、Hadoop 集群搭建、Hadoop 集群测试和 Hadoop 集群初步体验几个方面，带领大家从零开始搭建一个简单的 Hadoop 集群，并初步体验 Hadoop 集群的简单实用。

2.1 安装准备

Hadoop 可以安装在 Linux 系统或 Windows 系统上使用。由于 Linux 系统具备便捷性和稳定性，所以在实际开发过程中，更多的 Hadoop 集群是在 Linux 系统上运行的。本书对 Linux 系统上的 Hadoop 集群搭建以及使用进行讲解。

2.1.1 虚拟机安装

搭建 Hadoop 集群需要很多台机器，这在个人开发测试和学习时肯定是不切实际的。我们可以使用虚拟机软件在一台计算机中搭建出多个 Linux 虚拟机环境，来进行个人开发测试和学习。下面分步演示使用 VMware Workstation 虚拟软件工具进行 Linux 系统虚拟机安装配置的过程。

1. 创建虚拟机

（1）根据说明下载并安装 VMware Workstation 虚拟软件工具（此次演示使用 VMware Workstation 14 Pro 版本，具体安装步骤见附录），安装成功后打开 VMware Workstation，效果如图 2.1 所示。

图 2.1 VMware Workstation 界面

（2）在图 2.1 中，单击"创建新的虚拟机"选项，根据提示可以使用默认安装方式连续单击"下一步(N)"按钮。在进入"选择客户机操作系统"界面时，选择此次要安装的客户机操作系统 Linux，以及版本 CentOS 6 64 位，如图 2.2 所示。

（3）在图 2.2 中选择客户机操作系统后，单击"下一步(N)"按钮，进入"命名虚拟机"界面，自定义设置虚拟机名称（演示中设置虚拟机名称为 qf01）和安装位置，如图 2.3 所示。

图 2.2 选择客户机操作系统

图 2.3 命名虚拟机

（4）在图 2.3 中完成虚拟机命名后，单击"下一步(N)"按钮，进入"处理器配置"界面，根据个人计算机的硬件情况和使用需求，自定义配置处理器数量和每个处理器的内核数量，如图 2.4 所示。

（5）在图 2.4 中完成处理器配置后，单击"下一步(N)"按钮，进入"此虚拟机的内存"设置界面，再次根据个人计算机的物理内存进行合理分配。这里搭建的 qf01 虚拟机将作为 Hadoop 集群主节点，所以应分配较大内存，如图 2.5 所示。

图 2.4 处理器配置

图 2.5 虚拟机内存

（6）在图 2.5 中完成内存设置后，可以根据提示使用默认安装方式连续单击"下一步(N)"按钮。进入"指定磁盘容量"界面后，根据实际需要并结合计算机硬件情况合理选择最大磁盘大小（演示中使用默认值 20GB），如图 2.6 所示。

（7）在图 2.6 中完成磁盘容量设置后，根据提示使用默认安装方式连续单击"下一步(N)"按钮。在进入"已准备好创建虚拟机"界面时，可查看当前设置的要创建的虚拟机参数，确认无误后单击"完成"按钮，即可完成新建虚拟机的设置，如图 2.7 所示。

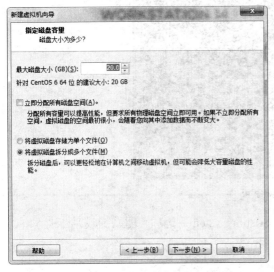

图 2.6 指定磁盘容量

图 2.7 创建完成

根据上述步骤和说明进行操作，便可以完成新建虚拟机的设置。接下来，对该虚拟机进行启动和初始化。

2. 虚拟机启动和初始化

（1）在创建成功的 qf01 虚拟机上单击鼠标右键，选择"设置"，然后选中设备中的"CD/DVD(IDE)"，勾选"使用 ISO 镜像文件"选项，单击"浏览（B）"按钮来指定 ISO 镜像文件的具体地址（此处根据前面操作系统的设置使用 CentOS 镜像文件初始化 Linux 系统），如图 2.8 所示。

图 2.8 挂载镜像

（2）设置完 ISO 镜像文件后，单击图 2.8 中的"确定"按钮，选择当前 qf01 主界面的"开启此虚拟机"选项，来启动 qf01 虚拟机，启动后效果如图 2.9 所示。

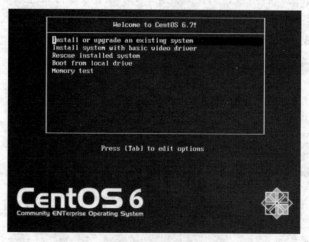

图 2.9 启动界面

（3）选择图 2.9 中第一条"Install or upgrade an existing system"选项，驱动加载完毕进入"Disc Found"界面，如图 2.10 所示。

图 2.10 "Disc Found"界面

（4）在图 2.10 中，按键盘→键切换至"Skip"选项并按 Enter 键进入 Cent OS 操作系统的初始化过程。首先单击界面的"Next"按钮进入系统语言设置界面，为了保证系统兼容性，通常会使用默认的"English(English)"选项（为了方便查看也可以选择"Chinese(Simplified)（中文（简体））"选项，如图 2.11 所示。

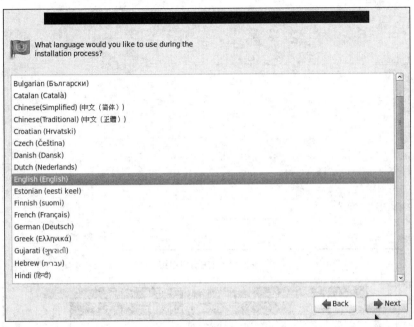

图 2.11　系统语言设置

（5）系统语言设置好后，使用系统默认配置连续单击界面的"Next（下一步）"按钮。当进入"Storage Device Warnig（存储设备警告）"界面时，按键盘←键切换至"Yes，discard any data（是，忽略所有数据）"选项并按 Enter 键，如图 2.12 所示。

图 2.12　存储设备警告

（6）执行完图 2.12 所示的存储设备确认后，会跳转到主机名（Hostname）设置界面，自定义设置该台虚拟机的主机名（演示中设置为 qf01），如图 2.13 所示。

（7）设置完主机名后，单击图 2.13 中的"Configure Network（网络配置）"选项，在弹窗中选择唯一的网卡"System eth0"，单击"Edit（编辑）"按钮，会出现一个网络配置的弹窗，在弹窗中勾选"Connect automatically（自动连接）"选项，并单击"Apply（应用）"按钮，如图 2.14 所示。

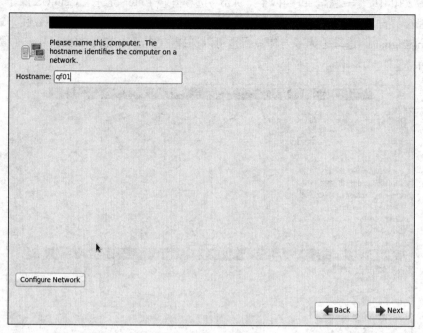

图 2.13　主机名设置

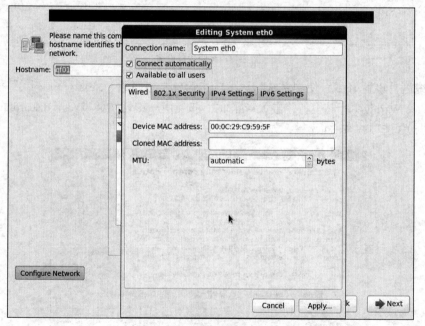

图 2.14　网络配置

（8）完成网络配置后，单击界面的"Next（下一步）"按钮，进行系统时区的选择。

（9）完成系统时区配置后，单击界面的"Next（下一步）"按钮，进入 root 用户密码设置界面，设置 root 用户密码，要求密码长度至少 6 个字符，如果密码强度较低可能出现提示窗口，直接单击"无论如何都用（Use Anyway）"按钮即可，如图 2.15 所示。

（10）完成系统 root 用户密码设置后，单击界面的"Next（下一步）"按钮，进入磁盘格式化界面，直接选择"Write changes to disk（将修改写入磁盘）"即可，如图 2.16 所示。

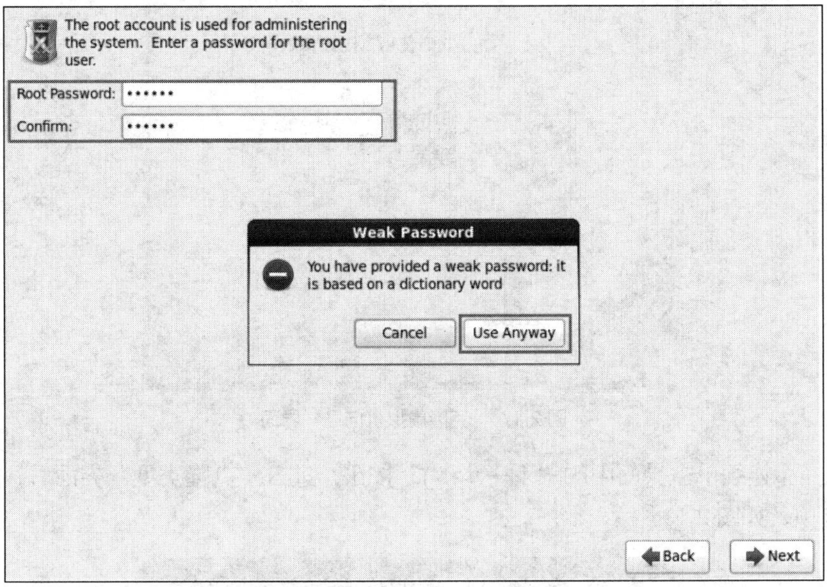

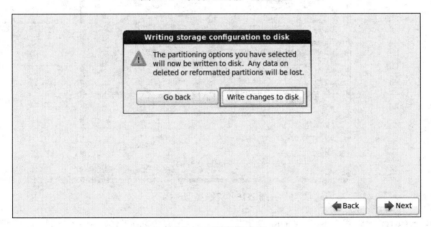

图 2.15　系统 root 用户密码设置

图 2.16　磁盘格式化

执行完上述操作后，该虚拟机进入磁盘格式化过程，稍等片刻后就会跳转到 CentOS 系统安装成功的界面。在安装成功界面，单击"Reboot（重启）"按钮进行重新启动，至此，就完成了 CentOS 虚拟机的安装。

2.1.2　虚拟机克隆

一台搭载 CentOS 镜像文件的 Linux 虚拟机已经安装成功，但是要搭建 Hadoop 集群，一台虚拟机远远不能满足需求，这时需要对已安装的虚拟机进行克隆。

克隆就是复制原始虚拟机的全部状态，克隆操作一旦完成，克隆的虚拟机就可以脱离原始虚拟机独立存在，而且在克隆虚拟机和原始虚拟机中的操作是相对独立的，不相互影响。

（1）关闭虚拟机 qf01，克隆只能在虚拟机关机状态下进行。

（2）鼠标右键单击虚拟机名称，选择"管理"，再选择"克隆"，进入"克隆虚拟机向导"界面，如图 2.17 所示。

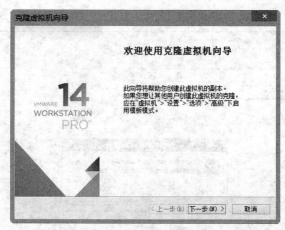

图 2.17 "克隆虚拟机向导"界面

（3）根据提示连续单击界面中的"下一步(N)"按钮，进入"克隆类型"界面后，选择"创建完整克隆(F)"选项，如图 2.18 所示。

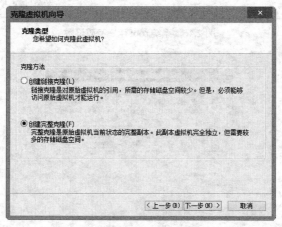

图 2.18 克隆类型选择

（4）在图 2.18 中，单击"下一步（N）"按钮，进入设置虚拟机名称和位置界面，如图 2.19 所示。设置虚拟机名称为 qf02，位置为 D:\hadoop\qf02，单击"完成"按钮。

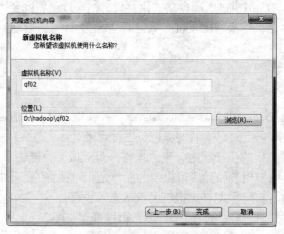

图 2.19 新虚拟机名称

在图 2.19 中，单击"完成"按钮，进入新虚拟机克隆过程，完成后在 VMware 的库列表中可以看到刚才克隆的虚拟机。

上面演示了一台虚拟机的克隆方法，如果想克隆多台虚拟机，重复上述操作即可。

2.1.3 Linux 系统网络配置

VMware Workstation 的虚拟网络类型主要有 3 种：桥接模式、NAT（Network Address Translation，网络地址转换）模式、仅主机模式。三者的用途各不相同。

（1）桥接模式可以将虚拟机直接连接到外部网络。
（2）NAT 模式可以与虚拟机共享主机的 IP 地址。
（3）仅主机模式可以在专用网络内连接虚拟机。

本书中使用虚拟网络的 NAT 模式，配置方式如下。

1. 主机名和 IP 映射配置

开启虚拟机 qf01，输入 root 用户的用户名和密码后进入虚拟机系统。然后，在终端窗口按照以下说明进行主机名和 IP 映射的配置。

（1）配置主机名，具体指令如下。

```
$ vi /etc/sysconfig/network
```

执行上述指令后，在打开的界面中对主机名进行重新编辑。在 Hadoop 集群搭建时，通常会将 qf01、qf02、qf03 主机名依次设置为 qf01、qf02 和 qf03。

（2）配置 IP 映射

配置 IP 映射，要明确目前虚拟机的 IP 地址和主机名，并且 IP 地址须在 VMware 虚拟网络 IP 地址范围之内。所以，要先查询可选的 IP 地址范围，再进行 IP 映射配置。

首先，单击 VMware 工具的"编辑"菜单下的"虚拟网络编辑器"菜单项，打开虚拟网络编辑器；选择"NAT 模式"类型的 VMnet8，单击"DHCP 设置(P)"按钮会出现一个 DHCP 设置弹窗，如图 2.20 所示。

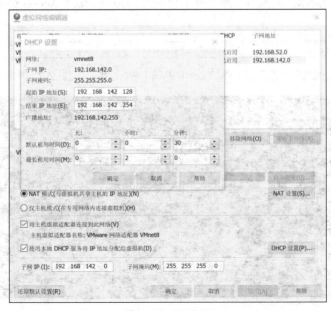

图 2.20　DHCP 设置

从图 2.20 中可以看出，此处 VMware 工具允许的虚拟机 IP 地址为 192.168.142.128～192.168.142.254（不同计算机网络可能不同）。因此，明确了要配置 IP 映射的 IP 地址可选范围（且不建议使用已用 IP 地址）。

然后，执行如下指令，对 IP 映射文件 hosts 进行编辑。

```
$ vi /etc/hosts
```

执行上述指令后，会打开一个 hosts 映射文件，为了保证后续相互关联的虚拟机能够通过主机名进行访问，自定义配置对应的 IP 和主机名映射，如图 2.21 所示。

图 2.21　IP 映射

从图 2.21 可以看出，此处将主机名 qf01、qf02、qf03 分别与 IP 地址 192.168.142.131、192.168.142.132 和 192.168.142.133 进行了匹配映射（将要搭建的集群主机都配置主机名和 IP 映射）。读者在进行 IP 映射配置时，可以根据自己的 DHCP 设置和主机名规划 IP 映射。

2. 网络参数配置

想要虚拟机能够正常使用，需要进行网络参数配置。

（1）修改虚拟机的网卡配置文件，配置网卡设备的 MAC（Media Access Control，介质访问控制）地址，具体指令如下。

```
$ vi /etc/udev/rules.d/70-persistent-net.rules
```

执行上述指令后，会打开当前虚拟机的网卡设备参数文件，如图 2.22 所示。

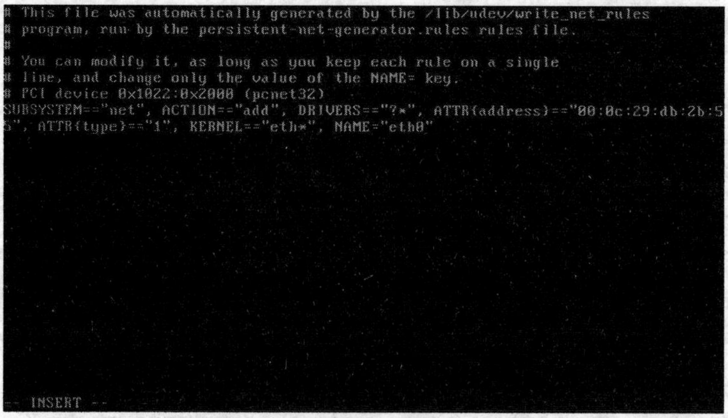

图 2.22　网卡配置

因为虚拟机 qf02 是克隆的，所以 qf02 中会有 eth0 和 eth1 两块网卡（qf01 只有一块网卡 eth0），这时删除多余的 eth1 网卡配置，只保留 eth0 一块网卡即可。操作方式：删除 eth0 网卡，将 eth1 网卡的参数 NAME="eth1"修改为 NAME="eth0"。

（2）修改 IP 地址文件，设置静态 IP，具体指令如下。

```
$ vi /etc/sysconfig/network-scripts/ifcfg-eth0
```

执行上述指令，打开虚拟机的 IP 地址配置界面，如图 2.23 所示。

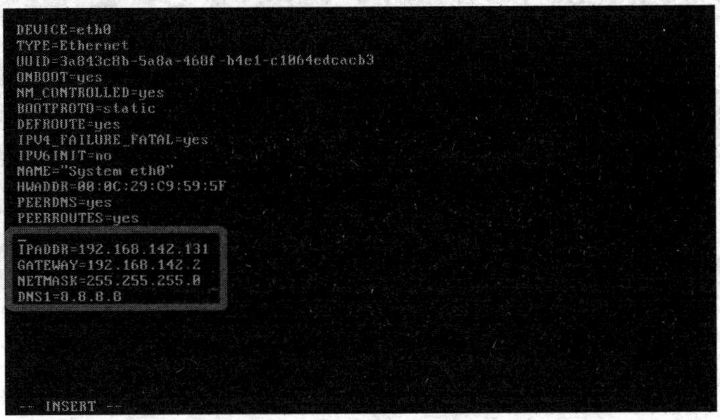

```
DEVICE=eth0
TYPE=Ethernet
UUID=3a843c8b-5a8a-468f-b4c1-c1064edcacb3
ONBOOT=yes
NM_CONTROLLED=yes
BOOTPROTO=static
DEFROUTE=yes
IPV4_FAILURE_FATAL=yes
IPV6INIT=no
NAME="System eth0"
HWADDR=00:0C:29:C9:59:5F
PEERDNS=yes
PEERROUTES=yes

IPADDR=192.168.142.131
GATEWAY=192.168.142.2
NETMASK=255.255.255.0
DNS1=8.8.8.8

-- INSERT --
```

图 2.23　IP 地址配置

在图 2.23 所示的 IP 地址配置界面，配置以下参数。

- ONBOOT=yes：表示启动这块网卡。
- BOOTPROTO=static：表示静态路由协议，可以保持 IP 固定。
- IPADDR：表示虚拟机的 IP 地址，这里设置的 IP 地址要与前面 IP 映射配置时的 IP 地址保持一致，否则无法通过主机名找到对应 IP。
- GATEWAY：表示虚拟机网关，一般是将 IP 地址最后一位数变成 2。
- NETMASK：表示虚拟机子网掩码，配置为 255.255.255.0。
- DNS1：表示域名解析器，此处采用 Google 提供的免费 DNS 服务器 8.8.8.8（也可以设置为 PC 端计算机对应的 DNS）。

3. **配置效果验证**

重启虚拟机使当前配置生效，这里可以使用 reboot 指令重启系统。系统重启完毕之后，通过 ifconfig 指令查看 IP 配置是否生效，如图 2.24 所示。

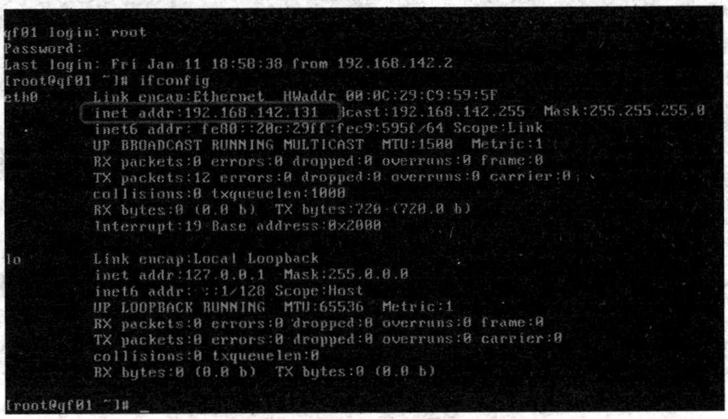

```
qf01 login: root
Password:
Last login: Fri Jan 11 18:58:38 from 192.168.142.2
[root@qf01 ~]# ifconfig
eth0      Link encap:Ethernet  HWaddr 00:0C:29:C9:59:5F
          inet addr:192.168.142.131  Bcast:192.168.142.255  Mask:255.255.255.0
          inet6 addr: fe80::20c:29ff:fec9:595f/64 Scope:Link
          UP BROADCAST RUNNING MULTICAST  MTU:1500  Metric:1
          RX packets:0 errors:0 dropped:0 overruns:0 frame:0
          TX packets:12 errors:0 dropped:0 overruns:0 carrier:0
          collisions:0 txqueuelen:1000
          RX bytes:0 (0.0 b)  TX bytes:720 (720.0 b)
          Interrupt:19 Base address:0x2000

lo        Link encap:Local Loopback
          inet addr:127.0.0.1  Mask:255.0.0.0
          inet6 addr: ::1/128 Scope:Host
          UP LOOPBACK RUNNING  MTU:65536  Metric:1
          RX packets:0 errors:0 dropped:0 overruns:0 frame:0
          TX packets:0 errors:0 dropped:0 overruns:0 carrier:0
          collisions:0 txqueuelen:0
          RX bytes:0 (0.0 b)  TX bytes:0 (0.0 b)

[root@qf01 ~]# _
```

图 2.24　查看 IP 配置

从图 2.24 中可以看出，主机 qf01 的 IP 地址已经设置为 192.168.142.131。执行 ping 192.168.142.132 指令检测集群通信是否正常，如图 2.25 所示。

```
lo        Link encap:Local Loopback
          inet addr:127.0.0.1  Mask:255.0.0.0
          inet6 addr: ::1/128 Scope:Host
          UP LOOPBACK RUNNING  MTU:65536  Metric:1
          RX packets:0 errors:0 dropped:0 overruns:0 frame:0
          TX packets:0 errors:0 dropped:0 overruns:0 carrier:0
          collisions:0 txqueuelen:0
          RX bytes:0 (0.0 b)  TX bytes:0 (0.0 b)

[root@qf01 ~]#
[root@qf01 ~]#
[root@qf01 ~]# ping 192.168.142.132
PING 192.168.142.132 (192.168.142.132) 56(84) bytes of data.
From 192.168.142.131 icmp_seq=1 Destination Host Unreachable
From 192.168.142.131 icmp_seq=2 Destination Host Unreachable
From 192.168.142.131 icmp_seq=3 Destination Host Unreachable
From 192.168.142.131 icmp_seq=5 Destination Host Unreachable
From 192.168.142.131 icmp_seq=6 Destination Host Unreachable
From 192.168.142.131 icmp_seq=7 Destination Host Unreachable
^C
--- 192.168.142.132 ping statistics ---
7 packets transmitted, 0 received, +6 errors, 100% packet loss, time 6970ms
pipe 3
[root@qf01 ~]#
```

图 2.25　检测集群通信

从图 2.25 中可以看到，虚拟机可以正常接收数据，并且延迟正常，说明网络连接正常。至此，当前虚拟机网络配置完毕。

2.1.4　SSH 服务配置

SSH 即 Secure Shell，是专为远程登录会话和其他网络服务提供安全保障的协议。

当 1 台计算机的某些软件频繁使用 SSH 协议远程连接其他计算机时，需要人工填写大量密码，影响工作效率。要避免这类情况出现，同时兼顾安全问题，可以设置 SSH 免密登录。

1. SSH 远程登录功能配置

在使用 SSH 服务之前，服务器必须安装并开启 SSH 服务。在 CentOS 系统中，执行 rpm -qa | grep ssh 指令来查看当前机器是否安装了 SSH 服务，然后使用 ps -e | grep sshd 指令来查看 SSH 服务是否已经正常开启，如图 2.26 所示。

```
          TX packets:0 errors:0 dropped:0 overruns:0 carrier:0
          collisions:0 txqueuelen:0
          RX bytes:0 (0.0 b)  TX bytes:0 (0.0 b)

[root@qf01 ~]#
[root@qf01 ~]#
[root@qf01 ~]# ping 192.168.142.132
PING 192.168.142.132 (192.168.142.132) 56(84) bytes of data.
From 192.168.142.131 icmp_seq=1 Destination Host Unreachable
From 192.168.142.131 icmp_seq=2 Destination Host Unreachable
From 192.168.142.131 icmp_seq=3 Destination Host Unreachable
From 192.168.142.131 icmp_seq=5 Destination Host Unreachable
From 192.168.142.131 icmp_seq=6 Destination Host Unreachable
From 192.168.142.131 icmp_seq=7 Destination Host Unreachable
^C
--- 192.168.142.132 ping statistics ---
7 packets transmitted, 0 received, +6 errors, 100% packet loss, time 6970ms
pipe 3
[root@qf01 ~]# rpm -qa | grep ssh
libssh2-1.4.2-1.el6_6.1.x86_64
openssh-5.3p1-111.el6.x86_64
openssh-server-5.3p1-111.el6.x86_64
[root@qf01 ~]# ps -e | grep sshd
  1516 ?        00:00:00 sshd
[root@qf01 ~]#
```

图 2.26　查看是否安装和开启 SSH 服务

从图 2.26 可以看出，CentOS 虚拟机已经默认安装并开启了 SSH 服务，不需要再进行额外的安装，就可以进行远程连接访问（如果没有安装，CentOS 系统下可以执行 yum install openssh-server 指令进行安装）。

若目标服务器已经安装 SSH 服务，并且支持远程连接访问，便可以通过一个远程连接工具来连接访问目标服务器。本书使用实际开发中常用的 MobaXterm 远程连接工具来演示远程服务器的连接和使用。

MobaXterm 是一款支持 SSH 的终端仿真程序，它能够在 Windows 操作系统上远程连接 Linux 服务器，执行操作。本书采用 MobaXterm 9.4 版本进行介绍说明（软件安装包见附录），下载安装完成后，按照以下步骤进行远程连接访问。

（1）打开 MobaXterm 远程连接工具，单击导航栏上的"Session"，然后单击"SSH"创建快速连接，并根据虚拟机的配置信息进行设置，如图 2.27 所示。

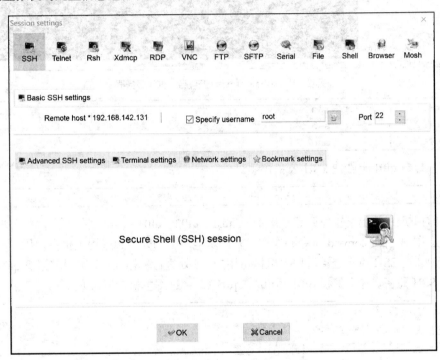

图 2.27　创建快速连接

在图 2.27 所示的快速连接设置中，主要是根据要连接的远程服务器设置了目标主机 192.168.142.131（qf01 虚拟机的 IP 地址）和登录用户 root，其他相关设置通常情况下使用默认值即可。

（2）在图 2.27 中单击"OK（确定）"按钮后，MobaXterm 远程连接工具就会自动连接到远程目标服务器，如图 2.28 所示。进入图 2.28 所示界面，就表示通过 MobaXterm 远程连接服务器成功，然后就可以像在虚拟机终端窗口中一样，在该工具客户端上操作虚拟机。

2. SSH 免密登录功能配置

如果 1 台计算机需要 SSH 远程登录其他计算机，就在这台计算机上配置 SSH 免密登录。SSH 免密登录是通过配置公钥和私钥（"公私钥"认证）来实现的。此处的 SSH 免密登录是指虚拟机 qf01 通过 SSH 免密登录虚拟机 qf02、qf03。

（1）分别删除虚拟机 qf01、qf02、qf03 的 ~/.ssh 目录。注意：在使用 rm -rf 命令时，要特别小心，避免误删文件。

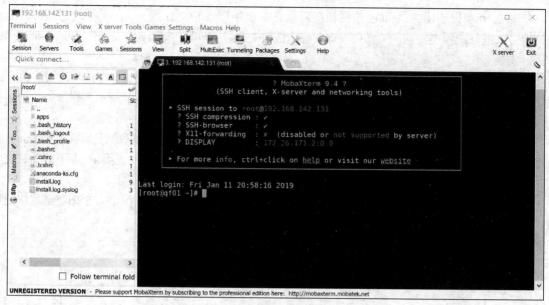

图 2.28　MobaXterm 远程连接到 qf01 服务器

```
[root@qf01 ~]# rm -rf .ssh
[root@qf02 ~]# rm -rf .ssh
[root@qf03 ~]# rm -rf .ssh
```

（2）在虚拟机 qf01 上新建 SSH 公私秘钥对。

```
[root@qf01 ~]# ssh-keygen -t rsa -P '' -f ~/.ssh/id_rsa
```

（3）在虚拟机 qf01 上配置免密登录虚拟机 qf01、qf02、qf03。

在实际工作中，ssh-copy-id 命令的作用：复制本地用户的公钥到远程主机指定用户的认证库中，实现本地用户通过 SSH 免密登录远程主机指定用户。此处进行模拟操作，实现虚拟机 qf01（本地 root 用户）通过 SSH 免密登录虚拟机 qf01、qf02、qf03（3 台远程主机的 root 用户）。

```
[root@qf01 ~]# ssh-copy-id root@qf01
[root@qf01 ~]# ssh-copy-id root@qf02
[root@qf01 ~]# ssh-copy-id root@qf03
```

提示：出现(yes/no)?，输入 yes，按 Enter 键。
（4）验证 SSH 免密登录是否配置成功。

```
[root@qf01 bin]# ssh qf02
```

出现如下内容，表明虚拟机 qf01 通过 SSH 成功登录虚拟机 qf02。

```
[root@qf02 ~]#
```

输入 exit，按 Enter 键，退回到虚拟机 qf01。

2.2　Linux 基本命令

Linux 基本命令的用法是我们必须要掌握的，它们与系统状态、目录、文件、网络等相关。熟练使用本节讲解的常用 Linux 命令，将为今后学习更复杂的命令打下良好基础。

2.2.1 系统工作命令

1. 查看系统信息

Linux 提供了查看系统信息的命令，如表 2.1 所示。

表 2.1　　　　　　　　　　　　　查看系统信息命令

命令	含义
uname -a	显示当前系统相关信息
uname -r	显示系统内核版本
uname -m	显示计算机类型
cat /proc/version	查看当前操作系统相关信息
cat /etc/redhat-release	查看当前操作系统发行版信息

2. 查看当前主机名

查看当前主机名，命令如下。

```
hostname
```

3. 查看网卡信息

查看网卡信息，命令如下。

```
ifconfig
```

4. 查看系统时间

查看系统时间，命令如下。

```
date
```

5. 查看进程状态

查看进程状态，命令如下。

```
ps -aux
```

6. 动态显示进程状态

动态显示进程状态，命令如下。

```
top
```

7. 以树状图显示进程间的关系

以树状图显示进程间的关系，命令如下。

```
pstree
```

8. 结束正在运行的指定进程

结束正在运行的指定进程，命令如下。

```
kill -9 进程 ID
```

9. 下载网络文件

下载网络文件，命令如下。

wget 下载地址

2.2.2 磁盘操作命令

1. 显示系统磁盘的空间用量

显示系统磁盘的空间用量命令，如表 2.2 所示。

表 2.2 　　　　　　　　　显示系统磁盘的空间用量命令

命令	含义
df -h	显示磁盘分区信息
fdisk -l	查看磁盘分区
fdisk /dev/sdb	管理磁盘分区
du -sh 目录或文件	查看目录或文件占用的空间大小

2. 用户与组操作命令

用户与组操作命令，如表 2.3 所示。

表 2.3 　　　　　　　　　用户与组操作命令

命令	含义
useradd xiaoqian	创建普通用户
passwd xiaoqian	设置用户密码
su – xiaoqian	切换用户
groupadd qf	创建用户组
gpasswd -a xiaoqian qf	将用户添加到组中
gpasswd -d xiaoqian qf	将用户从组中删除
groupdel qf	删除组
userdel xiaoqian	删除用户

2.2.3 目录与文件操作命令

1. 目录操作

目录操作包括创建目录、查看目录、切换目录、删除目录，具体命令如表 2.4 所示。

表 2.4 　　　　　　　　　目录操作命令

命令	含义
mkdir abc	创建一个空目录
mkdir -p aba/abb/abc	创建多级目录
pwd	查看当前所在目录
ls -l（可以简写为 ll）	查看目录与文件的属性
ls -a	查看隐藏的目录与文件
cd qf/aba/	切换目录
cd -	返回上次目录
rmdir abc	只删除空目录
rmdir -p abc/abd	连同上层空目录一起删除

2. 文件操作

文件操作包括创建文件、查看文件、复制文件、移动文件、删除文件,具体命令如表 2.5 所示。

表 2.5　　　　　　　　　　　　　　　文件操作命令

命令	含义
touch abc.txt	创建一个空白文件
echo hello > word.txt	新建 word.txt 文件,并写入内容 hello
ll abc.txt	查看文件信息
cat /etc/hosts	查看文件内容
more /etc/profile	逐页显示文件内容
head /etc/passwd	查看文件前几行的内容
tail /var/log/messages	查看文件后几行的内容
grep 'root' /etc/passwd	对文件内容进行过滤,搜索关键词
cp /tmp/file1.txt /opt	复制文件
cp -r /tmp/test01 /opt	复制目录
mv /opt/test01 /tmp	移动文件
rm linux.txt	删除文件
tar -cvf folder.tar file1.txt file2.txt	将多个文件打成一个包
tar -xvf folder.tar -C /home/xiaoqian	解包到指定目录
tar -zcvf file.tar.gz folder1 folder2	将多个文件打包并压缩
tar -zxvf /data3/data0.tar.gz -C /data2	将文件解包并解压缩到指定目录

2.2.4 权限操作命令

常用的 Linux 文件权限有如下几种。

```
444  -r--r--r--
644  -rw-r--r--
666  -rw-rw-rw-
755  -rwxr-xr-x
777  -rwxrwxrwx
```

以最后一行为例,可以看到-rwxrwxrwx,一共有 10 位。每位的含义如下。

(1)第 1 位表示文件或目录,-表示文件,还可能有 d 或 l 出现在-的位置,其中,d 表示目录,l 表示软链接文件或软链接目录。

(2)第 2、3、4 位表示文件所有者对文件或目录的权限,第 5、6、7 位表示同组用户对文件或目录的权限,第 8、9、10 位表示其他用户对文件或目录的权限。

在修改文件或目录权限时,文件所有者(user)用 u 表示,同组用户(group)用 g 表示,其他用户(other)用 o 表示,所有用户(all)用 a 表示。

(3)r 表示只读权限,w 表示写入权限,x 表示执行权限。

在修改文件或目录权限时,r 可以用数字 4 表示,w 可以用数字 2 表示,x 可以用数字 1 表示。

(4)+表示为指定的用户添加权限,-表示为指定的用户删除权限。具体示例如下。

chmod 777 1.txt	为文件的所有用户设置读写执行权限
chmod a+x 2.txt	为文件的所有用户设置执行权限
chmod u+x 1.txt	为文件的所有者添加执行权限
chmod o-w 1.txt	为文件的其他用户删除写入权限

2.3 Hadoop 集群搭建

在学习和个人开发测试时，可以安装多台 Linux 虚拟机来搭建 Hadoop 集群。我们已经学习了虚拟机的安装、网络配置以及 SSH 服务配置，下面就对 Hadoop 集群搭建进行详细讲解。

2.3.1 Hadoop 集群部署模式

在使用 Hadoop 之前，首先要了解 Hadoop 集群部署模式。Hadoop 集群部署模式分为三种，分别是单机模式（Standalone mode）、伪分布式模式（Pseudo-Distributed mode）、全分布式模式（Cluster mode），具体介绍如下。

（1）单机模式：Hadoop 的默认模式是单机模式。在不了解硬件安装环境的情况下，Hadoop 第一次解压其源码包时，会保守地选择最低配置，完全运行在本地。此时它不需要与其他节点进行交互，不使用 HDFS，也不加载任何 Hadoop 守护进程。单机模式不需要启动任何服务，一般只用于调试。

（2）伪分布式模式：全分布式模式的一个特例，Hadoop 的守护进程运行在一个节点上。伪分布式模式用于调试 Hadoop 分布式程序中的代码，以及验证程序是否准确执行。

（3）全分布式模式：Hadoop 的守护进程运行在由多个主机搭建的集群上，不同的节点担任不同角色，是真正的生产环境。在 Hadoop 集群中，服务器节点分为主节点（Master，1个）和从节点（Slave，多个），伪分布式模式是集群模式的特例，将主节点和从节点合二为一。

2.3.2 安装 JDK

因为 Hadoop 由 Java 语言开发，Hadoop 集群的使用同样依赖于 Java 环境，所以在搭建 Hadoop 集群前，需要先安装并配置 JDK（Java Development Kit，Java 开发工具包）。下面就在前面规划的 Hadoop 集群主节点 qf01 上分步骤演示如何安装和配置 JDK。

（1）将 JDK 安装包 jdk-8u121-linux-x64.tar.gz（见附录）放到/root/Downloads 目录下（直接拖进 MobaXterm 工具目录即可）。

（2）安装 JDK。命令如下。

```
[root@qf01 ~]# tar -zxvf jdk-8u121-linux-x64.tar.gz -C /usr/local/
```

（3）配置 JDK 环境变量。

安装完 JDK 后，还需要配置 JDK 环境变量。使用 vi /etc/profile 指令打开 profile 文件，在文件底部添加如下内容即可。

```
# 配置 JDK 系统环境变量
export JAVA_HOME=/usr/local/jdk1.8.0_121
export PATH=$PATH:$JAVA_HOME/bin
```

在/etc/profile 文件中配置完上述 JDK 环境变量后(注意 JDK 路径)，保存退出。然后，执行 source /etc/profile 指令方可使配置文件生效。

（4）JDK 环境验证。

在完成 JDK 的安装和配置后，为了检测安装效果，可以输入以下指令。

```
$ java -version
```

执行上述指令后,显示 JDK 版本信息,说明 JDK 安装和配置成功。

2.3.3 安装 Hadoop

Hadoop 是 Apache 软件基金会面向全球开源的产品之一(安装包见附录),读者可以从 Apache Hadoop 官方网站下载使用。本书以 Hadoop 2.7.3 版本为例。

(1)将安装包 hadoop-2.7.3.tar.gz(见附录)放到主节点 qf01 的/root/Downloads 目录下,并解压到/usr/local/目录下。命令如下。

```
[root@qf01 ~]# tar -zxvf /root/Downloads/hadoop-2.7.3.tar.gz -C /usr/local/
```

(2)打开文件/etc/profile,配置 Hadoop 环境变量。命令如下。

```
[root@qf01 ~]# vi /etc/profile
```

(3)编辑/etc/profile 时,依次按 G 键和 O 键,将光标移动到文件末尾,添加如下内容。编辑完成后,先按 Esc 键,后按 Shift+:组合键,输入 wq 保存退出。

```
# Hadoop environment variables
export JAVA_HOME=/usr/local/jdk1.8.0_121
export HADOOP_HOME=/usr/local/hadoop-2.7.3
export PATH=$PATH:$JAVA_HOME/bin:$HADOOP_HOME/bin:$HADOOP_HOME/sbin
```

(4)使配置文件生效。命令如下。

```
[root@qf01 ~]# source /etc/profile
```

(5)查看 Hadoop 版本,输入以下内容。

```
[root@qf01 ~]# hadoop version
```

输出如下内容即正确。

```
Hadoop 2.7.3
```

(6)查看 Hadoop 安装路径,输入如下内容。

```
[root@qf01 ~]# which hadoop
```

输出如下内容即正确。

```
/usr/local/hadoop-2.7.3/bin/hadoop
```

可以在 Hadoop 解压目录下通过 ll 指令查看 Hadoop 目录结构。下面对比较重要的目录内容及使用进行简单介绍。

- bin:Hadoop 相关服务(HDFS、YARN)的操作脚本存放在此目录下,但是通常使用的脚本在 sbin 目录下。
- etc:Hadoop 配置文件存放在此目录下,主要包含 core-site.xml、hdfs-site.xml、mapred-site.xml 等从 Hadoop 1.0 继承而来的配置文件和 yarn-site.xml 等 Hadoop 2.0 新增的配置文件。
- lib:Hadoop 对外提供的编程动态库和静态库存放在此目录下。
- sbin:Hadoop 管理脚本存放在此目录下,主要包含 HDFS 和 YARN 中各类服务的启动/关闭脚本。

2.3.4 Hadoop 集群配置

要在多台计算机上进行 Hadoop 集群搭建，还需要对相关配置文件进行修改，来保证集群服务协调运行。

Hadoop 集群搭建可能涉及的主要配置文件如表 2.6 所示。

表 2.6　　　　　　　　　　　　Hadoop 集群搭建相关配置文件

配置文件	功能描述
hadoop-env.sh	配置 Hadoop 运行所需的环境变量
yarn-env.sh	配置 YARN 运行所需的环境变量
core-site.xml	Hadoop 核心全局配置文件，可在其他配置文件中引用该文件
hdfs-site.xml	HDFS 配置文件，继承 core-site.xml 配置文件
mapred-site.xml	MapReduce 配置文件，继承 core-site.xml 配置文件
yarn-site.xml	YARN 配置文件，继承 core-site.xml 配置文件

在表 2.6 中，hodoop-env.sh 和 yarn-env.sh 配置文件指定 Hadoop 和 YARN 所需运行环境，hadoop-env.sh 配置文件用来保证 Hadoop 系统能够正常执行 HDFS 的守护进程 NameNode、SecondaryNameNode 和 DataNode；yarn-env.sh 配置文件用来保证 YARN 的守护进程 ResourceManager 和 NodeManager 能正常启动。其他 4 个配置文件用来设置集群运行参数，在这些配置文件中可以使用 Hadoop 默认配置文件中的参数来优化 Hadoop 集群，从而使集群更加稳定高效。

Hadoop 提供的默认配置文件 core-default.xml、hdfs-default.xml、mapred-default.xml 和 yarn-default.xml 中的参数很多，在此不再列举说明。具体使用时可以访问 Hadoop 官方文档，进入文档底部的 Configuration 部分进行学习和查看。

下面详细讲解 Hadoop 集群配置，具体步骤如下。

1. 配置 Hadoop 集群主节点

（1）修改 hadoop-env.sh 文件。

进入主节点 qf01 解压包下的 etc/hadoop/目录，用 vi hadoop-env.sh 指令打开其中的 hadoop-env.sh 文件，找到 JAVA_HOME 参数位置，进行如下修改（注意 JDK 路径）。

```
export JAVA_HOME=/usr/local/jdk1.8.0_121
```

上述配置文件中设置的 JDK 环境变量是 Hadoop 运行时需要的，使 Hadoop 启动时能够执行守护进程。

（2）在虚拟机 qf01 上，切换到/usr/local/hadoop-2.7.3/etc/hadoop/目录下。命令如下。

```
[root@qf01 ~]# cd /usr/local/hadoop-2.7.3/etc/hadoop
```

（3）配置 core-site.xml 文件。

core-site.xml 是 Hadoop 的核心配置文件，用于配置 HDFS 地址、端口号，以及临时文件目录。

① 打开 core-site.xml 文件。命令如下。

```
[root@qf01 hadoop]# vi core-site.xml
```

② 将 core-site.xml 文件中的内容替换为以下内容。

```
<configuration>
```

```
        <!-- 指定文件系统的名称-->
        <property>
            <name>fs.defaultFS</name>
            <value>hdfs://qf01:9000</value>
        </property>
        <!-- 配置 Hadoop 运行产生的临时数据存储目录 -->
        <property>
            <name>hadoop.tmp.dir</name>
            <value>/tmp/hadoop-qf01</value>
        </property>
        <!-- 配置操作 HDFS 的缓存大小 -->
        <property>
            <name>io.file.buffer.size</name>
            <value>4096</value>
        </property>
</configuration>
```

在上述文件中，配置了 HDFS 的主进程 NameNode 运行主机（Hadoop 集群的主节点），同时配置了 Hadoop 运行时生成数据的临时目录。

（4）配置 hdfs-site.xml 文件。

该文件用于设置 HDFS 的 NameNode 和 DataNode 两大进程。

① 打开 hdfs-site.xml 文件。命令如下。

```
[root@qf01 hadoop]# vi hdfs-site.xml
```

② 将 hdfs-site.xml 文件中的内容替换为以下内容。

```
<configuration>
    <!-- 配置 HDFS 块的副本数（全分布模式默认副本数是 3，最大副本数是 512） -->
    <property>
        <name>dfs.replication</name>
        <value>3</value>
    </property>
    <!-- 配置- secondary namenode 所在主机的 ip 和端口-->
    <property>
        <name>dfs.namenode.secondary.http-address</name>
        <value>qf02:50090</value>
    </property>
```

在上述配置文件中，HDFS 数据块的副本数量（默认值为 3，此处可省略）配置完成，并根据需要设置 SecondaryNameNode 所在主机的 HTTP 地址。

（5）配置 mapred-site.xml 文件。

此文件用于指定 MapReduce 运行框架，是 MapReduce 的核心配置文件。etc/hadoop/目录下没有该文件，需要先通过 cp mapred-site.xml.template mapred-site.xml 命令将文件复制并重命名为 mapred-site.xml，然后打开 mapred-site.xml 文件进行修改。

① 打开 mapred-site.xml 文件。命令如下。

```
[root@qf01 hadoop]# vi mapred-site.xml
```

② 将 mapred-site.xml 文件中的内容替换为以下内容。

```
<configuration>
    <!-- 指定 MapReduce 的运行框架 -->
```

```
    <property>
        <name>mapreduce.framework.name</name>
        <value>yarn</value>
    </property>
</configuration>
```

（6）配置 yarn-site.xml 文件。

在此文件中，指定 YARN 集群的管理者。

① 打开 yarn-site.xml 文件。命令如下。

```
[root@qf01 hadoop]# vi yarn-site.xml
```

② 将 yarn-site.xml 文件中的内容替换为以下内容。

```
<configuration>
    <!-- 指定启动 YARN 的 ResourceManager 服务的主机 -->
    <property>
        <name>yarn.resourcemanager.hostname</name>
        <value>qf01</value>
    </property>
    <!-- 配置 NodeManager 启动时加载 Shuffle 服务 -->
    <property>
        <name>yarn.nodemanager.aux-services</name>
        <value>mapreduce_shuffle</value>
    </property>
</configuration>
```

在上述配置文件中，配置 YARN 的主进程 ResourceManager 运行主机为 qf01，将 NodeManager 运行时的附属服务配置为 mapreduce_shuffle 才能正常运行 MapReduce 默认程序。

（7）设置从节点，即修改 slaves 文件。

此文件记录 Hadoop 集群所有从节点（HDFS 的 DataNode 和 YARN 的 NodeManager）的主机名，以配合使用脚本一键启动集群的从节点（需要保证关联节点配置了 SSH 免密登录）。

① 打开 slaves 文件。

```
[root@qf01 hadoop]# vi slaves
```

② 填写所有需要配置成从节点的主机名。具体做法是将 slaves 文件中的内容替换为以下内容。注意：每个主机名占一行。

```
qf01
qf02
qf03
```

2. 将集群主节点的配置文件分发到其他节点

完成 Hadoop 集群主节点 qf01 的配置后，还需要将系统环境配置文件、JDK 安装目录和 Hadoop 安装目录分发到其他节点 qf02 和 qf03 上，具体指令如下。

```
[root@qf01 ~]# scp /etc/profile qf02:/etc/profile
[root@qf01 ~]# scp /etc/profile qf03:/etc/profile
[root@qf01 ~]# scp -r /usr/local/hadoop-2.7.3 qf02:/usr/local/
[root@qf01 ~]# scp -r /usr/local/hadoop-2.7.3 qf03:/usr/local/
```

执行完上述所有指令后，还需要在 qf02 和 qf03 上分别执行 source /etc/profile 指令立即刷新配置文件。至此，整个集群所有节点都有了 Hadoop 运行所需要的环境和文件，Hadoop 集群安装配置完成。

2.4 Hadoop 集群测试

2.4.1 格式化文件系统

前面已经完成 Hadoop 集群的安装和配置，但还不能直接启动集群，因为在初次启动 HDFS 集群时，必须对主节点进行格式化处理，具体指令如下。

```
[root@qf01 ~]# hdfs namenode -format
```

或者

```
[root@qf01 ~]# hadoop namenode -format
```

执行上述任意一条命令均可以进行 Hadoop 集群格式化。执行格式化指令之后，出现 has been successfully formatted 信息，表明 HDFS 文件系统成功格式化，即可正式启动集群；否则，需要查看命令是否正确，或此前 Hadoop 集群的安装和配置是否正确。

此外需要注意的是，上述格式化命令只需要在 Hadoop 集群初次启动前执行一次即可，后续重复启动时不需要执行格式化。

2.4.2 启动和关闭 Hadoop 进程命令

Hadoop 集群的启动，需要启动其内部的两个集群框架，HDFS 集群和 YARN 集群。启动方式有单节点逐个启动和使用脚本一键启动两种。启动和关闭 Hadoop 进程的常用命令及含义，如表 2.7 所示。

表 2.7　启动和关闭 Hadoop 进程的常用命令及含义

命令	含义
start-dfs.sh	启动 Hadoop 的 HDFS 进程：NameNode、SecondaryNameNode、DataNode
stop-dfs.sh	关闭 Hadoop 的 HDFS 进程：NameNode、SecondaryNameNode、DataNode
hadoop-daemon.sh start namenode	单独启动某个节点的 NameNode 进程
hadoop-daemon.sh stop namenode	单独关闭某个节点的 NameNode 进程
hadoop-daemon.sh start datanode	单独启动某个节点的 DataNode 进程
hadoop-daemon.sh stop datanode	单独关闭某个节点的 DataNode 进程
hadoop-daemon.sh start namenode	启动所有节点的 NameNode 进程
hadoop-daemon.sh stop namenode	关闭所有节点的 NameNode 进程
hadoop-daemon.sh start datanode	启动所有节点的 DataNode 进程
hadoop-daemon.sh stop datanode	关闭所有节点的 DataNode 进程
hadoop-daemon.sh start secondarynamenode	单独启动 SecondaryNameNode 进程
hadoop-daemon.sh stop secondarynamenode	单独关闭 SecondaryNameNode 进程
start-yarn.sh	启动 Hadoop 的 YARN 进程：ResourceManager、NodeManager
stop-yarn.sh	关闭 Hadoop 的 YARN 进程：ResourceManager、NodeManager
yarn-daemon.sh start resourcemanager	单独启动 ResourceManager 进程
yarn-daemon.sh stop resourcemanager	单独关闭 ResourceManager 进程
yarn-daemon.sh start nodemanager	单独启动 NodeManager 进程
yarn-daemon.sh stop nodemanager	单独关闭 NodeManager 进程

2.4.3 启动和查看 Hadoop 进程

（1）在虚拟机 qf01 上，启动 Hadoop 进程。命令如下。

```
[root@qf01 hadoop]# start-dfs.sh
[root@qf01 hadoop]# start-yarn.sh
```

（2）查看 Hadoop 进程。

① 在虚拟机 qf01 中查看 Hadoop 进程。

```
[root@qf01 hadoop]# jps
3856 NameNode
5284 Jps
3974 DataNode
4472 NodeManager
4362 ResourceManager
```

提示：以 3856 NameNode 为例，3856 是进程 ID。

② 在虚拟机 qf02 中查看 Hadoop 进程。

```
[root@qf02 hadoop]# jps
3043 SecondaryNameNode
3398 DataNode
3575 NodeManager
4490 Jps
```

③ 在虚拟机 qf03 中查看 Hadoop 进程。

```
[root@qf03 hadoop]# jps
2978 DataNode
3221 NodeManager
3805 Jps
```

如果看到规划的 Hadoop 进程均已启动，则 Hadoop 全分布式集群搭建成功。

注意：由于只在 root 用户下搭建了 Hadoop 全分布式集群，再次启动虚拟机时，需要切换到 root 用户下，进行相关操作。

2.4.4 查看 Web 界面

Hadoop 集群正常启动后，默认开放 50070 和 8088 两个端口，分别用于监控 HDFS 集群和 YARN 集群。在本地操作系统的浏览器中输入集群服务的 IP 和对应的端口号即可访问。

为了后续方便查看，可以在本地宿主机的 hosts 文件（Windows 7 操作系统下路径为 C:\Windows\System32\drivers\etc\hosts）中添加集群服务的 IP 映射，具体内容示例如下（读者可以根据自己的集群构建进行相应的配置）。

```
192.168.142.131 qf01
192.168.142.132 qf02
192.168.143.133 qf03
```

要通过外部 Web 界面访问虚拟机服务，还需要对外开放 Hadoop 集群服务端口号。为了后续学习方便，这里直接关闭所有集群节点防火墙，具体操作如下。

首先，在所有集群节点上执行如下指令关闭防火墙。

```
[root@qf01 ~]# service iptables stop
```

然后，在所有集群节点上关闭防火墙开机启动，命令如下。

```
[root@qf01 ~]# chkconfig iptables off
```

执行完上述操作后，通过宿主机的浏览器分别访问 http://qf01:50070（集群服务 IP+端口号）和 http://qf01:8088，查看 HDFS 集群和 YARN 集群状态，效果分别如图 2.29 和图 2.30 所示。

图 2.29　HDFS 的 UI 界面

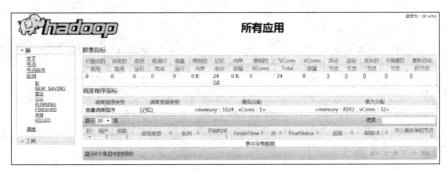

图 2.30　YARN 的 UI 界面

从图 2.29 和图 2.30 中可以看到，Hadoop 集群的 HDFS UI 界面和 YARN UI 界面通过 Web 界面均可以访问，并且页面显示正常，便于通过 Web 界面对集群状态进行管理和查看。

2.5　使用 Hadoop 集群

下面通过简单的案例，来初步了解 Hadoop 集群的使用。

（1）在集群主节点 qf01 上的/root/目录下，使用 vi word.txt 指令新建一个 word.txt 文本文件，并编写内容如下。

```
hadoop hive
hive hbase
flume sqoop
```

（2）上传 word.txt 到 HDFS 的 / 目录下。命令如下。

```
[root@qf01 ~]# hdfs dfs -put /word.txt /
```

（3）查看上传的 word.txt 文件。命令如下。

```
[root@qf01 ~]# hdfs dfs -cat /word.txt
```

（4）查看本地 word.txt 文件。命令如下。

```
[root@qf01 ~]# cat word.txt
```

也可以通过 Web 界面下载文件，用文本软件进行查看。如果文件的内容一致，表明 Hadoop 集群搭建成功。

2.6 本章小结

本章详细讲解了 Hadoop 集群的搭建过程。首先介绍集群搭建之前的准备工作，对虚拟机的安装和克隆、网络配置的步骤以及 SSH 免密登录配置方法进行了讲解。然后，详细地描述了 Hadoop 集群搭建过程，这是本章的重点内容，读者应亲手实践，在进行配置文件修改时，一定要认真细致。最后，我们对搭建好的 Hadoop 进行测试，并初步感受了 Hadoop 集群的使用。

2.7 习题

1. 填空题

（1）_____是 NameNode 的默认端口号，_____是 ResourceManager 的默认端口号。
（2）在主节点上启动所有 Hadoop 进程的命令是_____。
（3）查看网卡信息的命令是_____。
（4）显示磁盘分区信息的命令是_____。
（5）在 Linux 系统中，r 表示_____权限，w 表示_____权限，x 表示_____权限。

2. 选择题

（1）显示当前目录的命令是（　　）。
　　A. cd　　　　　B. pwd　　　　　C. who　　　　　D. ls
（2）Hadoop 的核心组件不包括（　　）。
　　A. HDFS　　　B. MapReduce　　C. YARN　　　　D. Common
（3）查看主机名称的命令是（　　）。
　　A. ifconfig　　 B. hostname　　　C. top　　　　　D. wget
（4）对文件重命名的命令是（　　）。
　　A. rm　　　　 B. mv　　　　　 C. move　　　　 D. cp

3. 思考题

（1）Hadoop 集群部署模式有几种？
（2）简述 MapReduce 和 YARN 之间的联系。

第3章 HDFS分布式文件系统

本章学习目标
- 掌握 HDFS 的架构和原理
- 掌握 HDFS 的 Shell 和 Java API 操作方法
- 了解 Hadoop 序列化
- 了解 Hadoop 小文件处理方式

Hadoop 的核心是 HDFS 和 MapReduce。HDFS 由 NDFS 系统演变而来，主要解决海量大数据存储的问题，也是目前分布式文件系统中应用比较广泛的一个。本章将带领大家理解和运用 HDFS 系统。

3.1 HDFS 简介

3.1.1 HDFS 的概念

HDFS（Hadoop Distributed File System，Hadoop 分布式文件系统）是一种通过网络实现文件在多台主机上进行分布式存储的文件系统。分布式存储比普通存储方式节省时间。

例如，现有 10 台计算机，每台计算机上有 1TB 的硬盘，如果将 Hadoop 安装在这 10 台计算机上，就可以使用 HDFS 进行分布式的文件存储。相当于登录到一台具有 10 TB 存储容量的大型机器。而用 HDFS 分布式文件存储方式在 10 台计算机上存储，显然比用普通方式在 1 台计算机上存储更节省时间，这就如同 3 个人吃 3 个苹果比 1 个人吃 3 个苹果要快。

1. NameNode

NameNode（名称节点）管理文件系统的命名空间。它负责维护文件系统树及树内所有的文件和目录。这些信息以两个文件（命名空间镜像文件和编辑日志文件）的形式永久保存在本地磁盘上。同时 NameNode 也记录着每个文件中各个块所在的数据节点信息，但它并不永久保存块的位置信息，因为这些信息在系统启动时由数据节点重建。

2. DataNode

DataNode（数据节点）是 HDFS 实例中在单独机器上运行的软件，Hadoop 集群包含一个 NameNode 和大量的 DataNode。一般情况下 DataNode 以机架的形式组织，机架通过一个交换机把所有的系统连接起来。Hadoop 的一个假设是：机架内部节点之间的传输速度要快于机架间的传输速度。

DataNode 响应来自 HDFS 客户机的读写请求。它们还响应来自 NameNode 的创建、删除和复制块的命令。NameNode 依赖来自每个 DataNode 的定期心跳（Heartbeat）消息。每条消息都包含一个块报告，NameNode 可以根据这个报告验证块映射和其他文件系统元数据。如果 DataNode 不发送心跳消息，NameNode 将采取修复措施，重新复制该节点上丢失的块。

3.1.2 HDFS 数据的存储和读取方式

（1）对于大文件的存储，HDFS 采用分割的方式。HDFS 将大文件分割到既定的存储块（Block）中进行存储，并通过本地设定的任务节点进行预处理。

（2）对于大量小文件的存储，HDFS 采用普通的编码与压缩方式。在实际工作中，更多时候需要存储大量的小文件。

（3）对于普通文件的读取，HDFS 通常采用分批次的方式。

（4）对于大量数据的读取，HDFS 采用集中式。存储时的优化使得读取能够在一个连续的区域内进行，节省读取数据的时间。

（5）对于少量数据的随机读取，HDFS 一般采用按序读取的方式，即先把少量的随机读取操作合并，然后按顺序读取数据。

3.1.3 HDFS 的特点

1. HDFS 的优点

（1）成本低。HDFS 可以部署在价格低廉的硬件上，成本较低。例如，只要有台式机或笔记本电脑，就可以部署 HDFS。

（2）高容错。HDFS 利用众多服务器实现分布式存储，每个数据文件都有 2 个冗余备份，也就是每个数据文件被存储 3 次。如果存储数据的某个服务器发生了故障，数据还有 2 个备份，因此，HDFS 具有高容错的特性，允许服务器发生故障。

（3）高吞吐量。吞吐量是指单位时间内完成的工作量。HDFS 实现了并行处理海量数据，大大缩短了处理时间，从而实现了高吞吐量。

（4）存储数据种类多。HDFS 可以存储任何类型的数据，如结构化的数据、非结构化的数据、半结构化的数据。

（5）可移植。HDFS 可以实现不同平台之间的移植。

2. HDFS 的缺点

（1）高延时。HDFS 不适用于对延时敏感的数据访问。

（2）不适合小文件存取场景。对 Hadoop 系统而言，小文件通常指远小于 HDFS 的数据块大小（128MB）的文件，由于每个文件都会产生各自的元数据，Hadoop 通过 NameNode 来存储这些信息，若小文件过多，容易导致 NameNode 存储出现瓶颈。

（3）不适合并发写入。HDFS 目前不支持并发多用户的写操作，写操作只能在文件末尾追加数据。

3.2 HDFS 存储架构和数据读写流程

3.2.1 HDFS 的存储架构

对于 HDFS 架构来说，一个 HDFS 基本集群的节点主要包括 NameNode、DataNode、SecondaryNameNode。

HDFS 采用主/从模式架构，如图 3.1 所示。

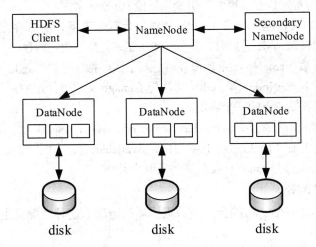

图 3.1　HDFS 架构图

1. HDFS 的 Client、NameNode、DataNode

在图 3.1 中，Client 是客户端，NameNode 是名称节点，SecondaryNameNode 是辅助名称节点，DataNode 是数据节点。下面分别介绍各部分的功能。

（1）Client。Client 主要负责切分文件，与各节点进行交互。

切分文件是把文件分割成数据块（Block）。数据块默认大小是 128MB，每个数据块有多个副本存储在不同的机器上，副本数可在文件生成时指定（默认有 3 个副本）。

交互包括与 NameNode 和 DataNode 进行交互，Client 从 NameNode 获取文件的元数据，从 DataNode 读取数据并向 DataNode 写入数据。

元数据是描述数据属性的数据。数据属性包括存储位置、格式、权限、大小、历史版本等。

（2）NameNode。NameNode 是 HDFS 架构中的主节点（Master）。HDFS 架构中只有一个 NameNode。NameNode 主要负责管理 HDFS 的元数据，配置副本的放置策略，处理 Client 请求。

HDFS 副本的放置策略称为机架感知策略。以默认的 3 个副本为例，具体策略如下。

① 第一个副本放在本地机架的一个节点上。
② 第二个副本放在同一机架的另一个节点（随机选择）上。
③ 第三个副本放在不同机架的节点上。

如果还有更多的副本就随机放在集群的其他节点上。

这种策略减少了机架间的数据传输，从而提高了写操作的效率。机架故障的可能性远小于节点故障的可能性，因此这个策略保证了数据的可靠性。

为了降低整体的带宽消耗和读取延时，HDFS 会让程序尽量读取离它最近的副本。如果读取程序的同一个机架上有一个副本，那么就读取该副本；如果一个 Hadoop 集群跨越多个数据中心，那么优先读取本地数据中心的副本。

（3）DataNode。DataNode 是 HDFS 架构中的从节点（Slave）。DataNode 负责存储 Client 发来的数据块，执行数据块的读写操作，汇报存储信息给 NameNode。

（4）fsimage 和 edits。fsimage（镜像文件）和 edits（日志文件）是 NameNode 中两个很重要的文件。fsimage 是元数据镜像文件，内容是 NameNode 启动时对整个文件系统的快照（备份）。edits 是元数据操作日志，内容是每次保存 fsimage 之后至下次保存 fsimage 之前进行的所有 HDFS 操作。

（5）SecondaryNameNode。SecondaryNameNode 主要用来进行镜像备份，对 NameNode 中的 edits 文件与 fsimage 文件进行定期合并。镜像备份是指备份 fsimage 文件。

随着 HDFS 操作的次数越来越多，edits 文件也会越来越多，所占存储空间越来越大。如果 NameNode 出现故障，NameNode 重启时，会先把 fsimage 文件加载到内存中，然后合并 edits 文件，edits 文件占空间大会导致 NameNode 重启很耗时。对于实际操作来说，这是一个比较麻烦的问题。SecondaryNameNode 对 NameNode 中的 edits 文件与 fsimage 文件进行定期合并，很好地解决了这一问题，这也是 HDFS 高容错特点的一个表现。

合并 edits 文件和 fsimage 文件，SecondaryNameNode 每隔 1 小时执行 1 次，可能会出现数据丢失问题。目前大部分企业采用 Hadoop 的 HA（High Available，高可用）模式下备用的 NameNode，对 NameNode 中的 edits 文件与 fsimage 文件进行实时合并。

2. HDFS 数据损坏的处理

在 HDFS 架构使用过程中，可能会出现数据损坏的情况。这种情况发生时 HDFS 的处理步骤如下。

（1）DataNode 读取 Block 时，会计算 Checksum（校验和）。

（2）如果计算出的 Checksum 与 Block 创建时值不一样，说明该 Block 已经损坏。

（3）Client 读取其他 DataNode 上的 Block。

（4）NameNode 标记该 Block 已经损坏，然后复制 Block 达到预先设置的文件副本数。

（5）DataNode 在新文件创建三周后验证其 Checksum。

3.2.2 HDFS 的数据读写流程

HDFS 的数据读取流程，如图 3.2 所示。

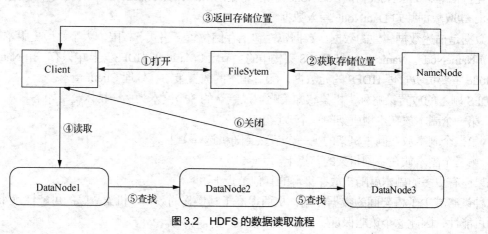

图 3.2 HDFS 的数据读取流程

下面对这一过程进行具体讲解。

（1）Client（客户端）通过调用 FileSystem 对象的 get() 方法打开需要读取的文件，这是一个常见的 HDFS 读取实例。

（2）FileSystem 通过远程过程调用协议调用 NameNode 的元数据来确定文件的前几个 Block 在 DataNode 上的存放位置，以便加快读取速度。接下来，DataNode 按照机架距离进行排序，如果 Client 本身就是一个 DataNode，那么优先从本地 DataNode 节点读取数据。HDFS 实例做完以上工作后，返回一个 FSDataInputStream（输入流）给 Client，让其从 FSDataInputStream 中读取数据。接下来，FSDataInputStream 封装一个 DFSInputStream 对象，用来管理 DataNode 和 NameNode 的输入或输出（Input 或 Output，I/O）。

（3）NameNode 向 Client 返回一个包含数据信息的地址，Client 根据地址创建一个 FSDataInputStream 开始对数据进行读取。

（4）FSDataInputStream 根据前几个 Block 在 DataNode 的存放位置，连接到最近的 DataNode。Client 反复调用 read()方法，从 DataNode 读取数据。

（5）当读到 Block 的结尾时，FSDataInputStream 会关闭当前 DataNode 的连接，查找能够读取下一个 Block 的最近的 DataNode。

（6）读取完成后，调用 close()方法，关闭 FSDataInputStream。

以上是 HDFS 的数据读取流程。

如果在 HDFS 数据读取期间，Client 与 DataNode 间的通信发生错误，Client 会寻找下一个最近的 DataNode。Client 记录发生错误的 DataNode，以后不再读取该 DataNode 上的数据。

HDFS 的数据写入流程，如图 3.3 所示。

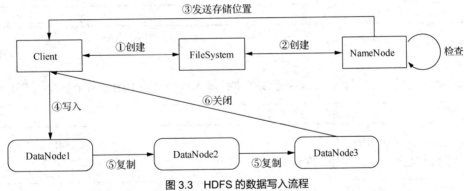

图 3.3 HDFS 的数据写入流程

下面对这一过程进行具体讲解。

（1）Client 通过调用 FileSystem 对象的 create()方法来请求创建文件。

（2）FileSystem 通过对 NameNode 发出远程请求，在 NameNode 里面创建一个新的文件。FileSystem 给 Client 返回一个 FSDataOutputStream（输出流）用来写入数据。FSDataOutputStream 封装一个 DFSOutputStream 对象用于 DataNode 与 NameNode 间的通信。

（3）FSDataOutputStream 把需要写入的数据分成 Packet（数据包），由 DataStreamer 来读取。DataStreamer 的职责是让 NameNode 给合适的 DataNode 分配 Block 来备份数据。假设有 3 个 DataNode，DataStreamer 将数据写入第一个 DataNode 后，第一个 DataNode 将 Packet 复制到第二个 DataNode，然后第二个 DataNode 将 Packet 复制到第三个 DataNode。

（4）FSDataOutputStream 维护了一个内部关于 Packet 的队列，其中存放等待被 DataNode 确认无误的 Packet 的信息。这个队列称为等待队列。当且仅当 Packet 被所有 DataNode 确认无误时，一个 Packet 才被移出本队列。

（5）Client 完成数据的写入后，会调用数据流的 close()方法，关闭数据流。Client 再将写入完成的信息反馈给 NameNode。

以上是 HDFS 的数据写入流程。如果写入数据时，DataNode 发生错误，其处理过程如下。

（1）发现错误后，关闭数据流，将没有被确认的数据放到数据队列的开头。当前的 Block 被赋予一个新的标识，发给 NameNode。在损坏的 DataNode 恢复后，删除这个没有被完成的 Block。

（2）移除损坏的 DataNode。NameNode 检测到一个 Block 的信息还没有被复制完成，转而在其他的 DataNode 上安排复制。接下来，Block 写入操作见 HDFS 的数据写入流程。

以上对 HDFS 数据的读取和写入流程进行了讲解。了解这些基本原理，对于深入学习和应用 HDFS 非常重要。

3.3 HDFS 的 Shell 命令

HDFS 提供了 Shell 命令来操作 HDFS，这些 Shell 命令和对应的 Linux Shell 命令相似，这样设计有利于使用者快速学会操作 HDFS。

在使用 Shell 命令操作 HDFS 之前，需要先启动 Hadoop。

```
[root@qf01 ~]# start-dfs.sh
```

HDFS Shell 命令的一般格式如下。

```
hdfs dfs -cmd 参数
```

HDFS 常见的 Shell 命令如下。

1. put

（1）put 命令可以上传本地（CentOS 系统）文件或本地目录到 HDFS 目录，使用方法如下。

```
hdfs dfs -put 本地文件或本地目录 HDFS 目录
```

（2）通过示例演示其用法。

① 新建本地文件 1.txt，并写入内容 hello。

```
[root@qf01 ~]# echo hello > 1.txt
```

② 把本地文件 1.txt 上传到 HDFS 的根目录下。

```
[root@qf01 ~]# hdfs dfs -put 1.txt /
```

③ 新建本地目录 data，把目录 data 上传到 HDFS 的根目录下。

```
[root@qf01 ~]# mkdir data
[root@qf01 ~]# hdfs dfs -put data /
```

2. ls

（1）ls 命令可以列出 HDFS 文件或目录，使用方法如下。

```
hdfs dfs -ls HDFS 文件/HDFS 目录
```

（2）通过示例演示其用法。

```
[root@qf01 ~]# hdfs dfs -ls /
```

```
-rw-r--r--   3 root supergroup          6 2018-07-17 11:50 /1.txt
drwxr-xr-x   - root supergroup          0 2018-07-17 12:00 /data
```

3. cat

（1）cat 命令可以查看 HDFS 文件的内容，使用方法如下。

```
hdfs dfs -cat HDFS 文件
```

（2）通过示例演示其用法。

```
[root@qf01 ~]# hdfs dfs -cat /1.txt
hello
```

4. get

（1）get 命令可以下载 HDFS 文件或目录到本地，使用方法如下。

```
hdfs dfs -get HDFS 文件或目录 本地目录
```

（2）通过示例演示其用法。

```
[root@qf01 ~]# rm -f 1.txt
[root@qf01 ~]# hdfs dfs -get /1.txt ~
[root@qf01 ~]# cat 1.txt
hello
```

5. 其他常用命令

命令	说明
hdfs dfs -mkdir /aaa	在 HDFS 的根目录下创建目录 aaa
hdfs dfs -cp /1.txt /2.txt	复制 HDFS 的/1.txt 重命名为根目录下的/2.txt
hdfs dfs -mv /1.txt /aaa/	移动 HDFS 的/1.txt 到/aaa 目录下
hdfs dfs -mv /2.txt /3.txt	重命名 HDFS 根目录下的/2.txt 为/3.txt
hdfs dfs -rm /3.txt	删除 HDFS 的文件/3.txt
hdfs dfs -rm -r /data	删除 HDFS 的目录/data
hdfs dfs -chmod 777 /aaa/1.txt	为 HDFS 的/aaa/1.txt 文件的所有用户设置读写执行权限

3.4　Java 程序操作 HDFS

除了可以使用 HDFS Shell 命令操作 HDFS 外，还可以通过 Java 程序操作 HDFS。在实际工作中，大多数情况下是通过程序来操作 HDFS 的。

3.4.1　HDFS Java API 概述

Java 程序通过 Hadoop 提供的文件操作类可以进行读写 HDFS 文件、上传 HDFS 文件等操作。这些文件操作类都在 org.apache.hadoop.fs 包中（详见 Hadoop 官方网站的 Java API 文档）。

API（Application Programming Interface，应用程序编程接口）是一些预先定义的函数。API 使应用程序与开发人员获得了一种重要能力：无须访问源码或理解内部工作机制细节，即可访问一组例程。

3.4.2　使用 Java API 操作 HDFS

本节将通过 Java API 来演示如何操作 HDFS，包括文件上传、下载及目录操作。

1. 配置开发环境

（1）配置 Java 环境。

在 Windows 系统下配置 Java 环境，也就是安装 JDK，记住 JDK 安装路径。这里用的 JDK 版本是 1.8.0_121，安装包见附录（jdk-8u121-windows-x64.exe）。

（2）安装 IntelliJ IDEA（安装包见附录），导入配置文件 Settings，添加 Maven 依赖（配置 pom.xml 文件）。

① Maven 简介。Maven 是通过项目对象模型文件 pom.xml 来管理项目的构建、报告和文档的工具。简言之，Maven 是一种项目管理工具。

pom.xml 文件是 Maven 的基本单元。Maven 最强大的功能是通过 pom.xml 文件自动下载项目依赖库。

② 配置文件 Settings。用户在熟练使用 IntelliJ IDEA 后，可以对 IntelliJ IDEA 进行个性化配置。本书直接提供了 Settings 文件，有利于缩短前期的探索时间。

导入配置文件 Settings 的方法：单击 IntelliJ IDEA 菜单栏中的"File"，在下拉列表中单击"Import Settings..."选项，选择 Settings 文件的存放位置，单击"OK"即可。

2. IntelliJ IDEA 新建项目

IntelliJ IDEA 新建项目（Project）步骤如下。

（1）单击 IntelliJ IDEA 菜单栏中的"File"，在下拉列表中单击"New"选项，再在出现的列表中单击"Project..."选项（见图 3.4），进入新建项目对话框，如图 3.5 所示。

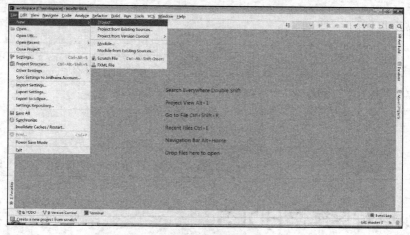

图 3.4　新建项目菜单

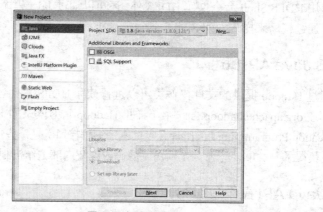

图 3.5　新建项目

（2）在图 3.5 的"Project SDK"一栏中，单击"New"按钮，出现选择 JDK 安装路径对话框，如图 3.6 所示。

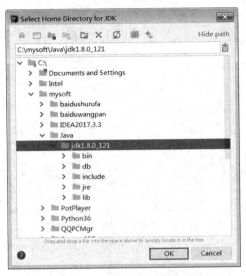

图 3.6　选择 JDK 安装路径

（3）在图 3.6 中，单击"jdk1.8.0_121"后，单击"OK"按钮，回到新建项目对话框；单击"Next"按钮，进入询问是否从模板中创建项目对话框，如图 3.7 所示。

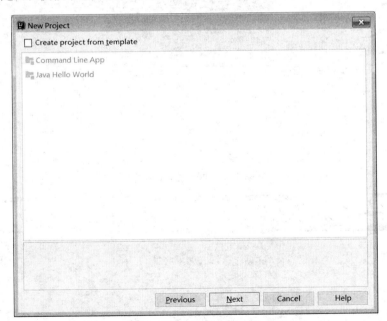

图 3.7　询问是否从模板中创建项目

（4）在图 3.7 中，单击"Next"按钮，进入填写项目名称和存放位置对话框，如图 3.8 所示。

（5）在图 3.8 中，在"Project name"栏中填写"testHadoop"，在"Project location"栏中填写"D:\IdeaProjects\testHadoop"，单击"Finish"按钮，出现询问是否创建目录对话框，如图 3.9 所示。

（6）在图 3.9 中，单击"OK"按钮，进入新建项目完成界面，如图 3.10 所示。

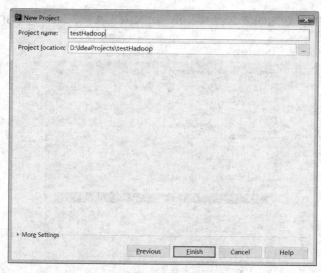

图 3.8 填写项目名称和存放位置

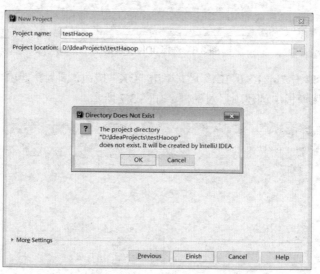

图 3.9 询问是否创建目录

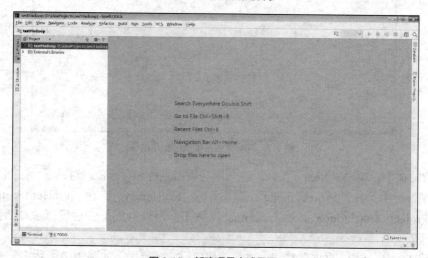

图 3.10 新建项目完成界面

至此，IntelliJ IDEA 新建项目完成。

3. IntelliJ IDEA 新建模块

IntelliJ IDEA 新建模块（Module）步骤如下。

（1）在图 3.10 中，右键单击左侧"Project"栏中的项目名"testHadoop"，在出现的列表中单击"New"选项，再在出现的列表中单击"Module"选项（见图 3.11），进入新建模块对话框，如图 3.12 所示。

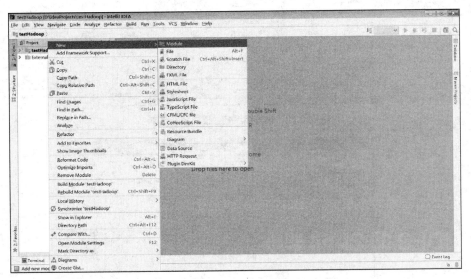

图 3.11　新建模块菜单

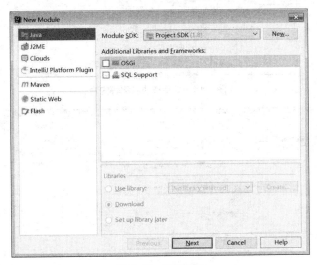

图 3.12　新建模块

（2）在图 3.12 中，单击"Maven"，在"Module SDK"一栏中，单击"New"按钮，选择 JDK 的安装路径（不再赘述）。单击"Next"按钮，进入填写组织名、模块名、版本对话框，如图 3.13 所示。

（3）在图 3.13 中的"GroupId"栏填写组织名"com.qf"，在"ArtifactId"栏填写模块名"testHDFS"，单击"Next"按钮，进入填写模块名称等内容的对话框，如图 3.14 所示。

（4）在图 3.14 中，单击"Finish"按钮，出现待填写的 pom.xml 文件，如图 3.15 所示。

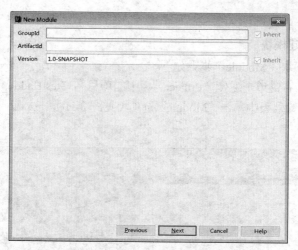

图 3.13 填写组织名、模块名、版本

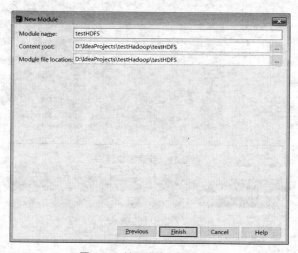

图 3.14 填写模块名称等内容

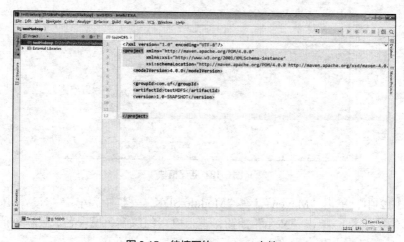

图 3.15 待填写的 pom.xml 文件

（5）修改 pom.xml 文件内容如下。

```
<?xml version="1.0" encoding="UTF-8"?>
<project xmlns="http://maven.apache.org/POM/4.0.0"
```

```xml
        xmlns:xsi="http://www.w3.org/2001/XMLSchema-instance"
        xsi:schemaLocation="http://maven.apache.org/POM/4.0.0
        http://maven.apache.org/xsd/maven-4.0.0.xsd">
    <modelVersion>4.0.0</modelVersion>
    <groupId>com.qf</groupId>                <!-- 组织名 -->
    <artifactId>testHDFS</artifactId>        <!-- 项目名 -->
    <version>1.0-SNAPSHOT</version>          <!-- 版本 -->
    <dependencies>
        <!-- Hadoop 客户端依赖，该依赖包含 HDFS 的相关依赖 -->
        <dependency>
            <groupId>org.apache.hadoop</groupId>
            <artifactId>hadoop-client</artifactId>
            <version>2.7.3</version>
        </dependency>
        <!-- 单元测试的依赖 -->
        <dependency>
            <groupId>junit</groupId>
            <artifactId>junit</artifactId>
            <version>4.12</version>
        </dependency>
    </dependencies>
</project>
```

至此，IntelliJ IDEA 新建模块完成。

4. 编写 Java 程序

（1）新建 Java OperateHDFS 类。在图 3.15 中，依次双击"Project"树状目录中的"testHadoop""testHDFS""src""main"目录，右键单击出现的"java"目录，在出现的列表中依次单击"New""Java Class"选项，在出现的创建新类对话框中的"Name"栏输入"com.qf.OperateHDFS"，单击"OK"按钮，出现程序编写界面，如图 3.16 所示。

图 3.16　程序编写界面

（2）通过 Java 程序操作 HDFS。

接下来演示使用 Java 程序实现三个功能：将数据写入 HDFS 文件，读取 HDFS 文件的数据，从 Windows 上传文件到 HDFS。

【例 3-1】OperateHDFS.java

```
1   public class OperateHDFS {
```

```java
2       @Test   //单元测试
3       /**
4        * 1.将数据写入 HDFS 文件
5        */
6       public void writeToHDFS() throws IOException {
7           //新建配置文件对象
8           Configuration conf = new Configuration();
9           //给配置文件设置 HDFS 文件的默认入口
10          conf.set("fs.defaultFS", "hdfs://192.168.142.131:9000");
11          //通过传入的配置参数得到 FileSystem
12          FileSystem fs = FileSystem.get(conf);
13          //获取 HDFS 上/1.txt 的绝对路径,注意: 一定要确保 HDFS 上/1.txt 是存在的,否则程序会报错
14          Path p = new Path("/1.txt");
15          //FileSystem 通过 create()方法获得输出流（FSDataOutputStream）
16          FSDataOutputStream fos = fs.create(p, true, 1024);
17          //通过输出流将内容写入 1.txt 文件
18          fos.write("helloworld".getBytes());
19          //关闭输出流
20          fos.close();
21      }
22      @Test
23      /**
24       * 2.读取 HDFS 文件
25       */
26      public void readHDFSFile() throws IOException {
27          //创建配置对象
28          Configuration conf = new Configuration();
29          //设置 HDFS 文件系统的网络地址和端口号
30          conf.set("fs.defaultFS", "hdfs://192.168.142.131:9000");
31          //通过配置获取文件系统
32          FileSystem fs = FileSystem.get(conf);
33          //在 HDFS 上的根目录下新建文件 1.txt
34          Path p = new Path("/1.txt");
35          //通过 FileSystem 的 open()方法获得数据输入流
36          FSDataInputStream fis = fs.open(p);
37          //分配 1024 个字节的内存给 buf（分配 1024 个字节的缓冲区）
38          byte[] buf = new byte[1024];
39          int len = 0;
40          //循环读取文件内容到缓冲区中，读到文件末尾结束（结束标识为-1）
41          while ((len = fis.read(buf)) != -1) {
42              //输出读取的文件内容到控制台
43              System.out.print(new String(buf, 0, len));
44          }
45      }
46      @Test
47      /**
48       * 3.从 Windows 上传文件到 HDFS
49       */
50      public void putFile() throws IOException {
51          Configuration conf = new Configuration();
52          conf.set("fs.defaultFS", "hdfs://192.168.142.131:9000");
```

```
53              FileSystem fs = FileSystem.get(conf);
54              //文件上传到HDFS上的位置
55              Path p = new Path("/");
56              //待上传文件1.sh在Windows系统中的绝对路径,此处需要提前在Windows系统中D盘下新建1.sh
文件,并写入Hello Xiaoqian
57              Path p2 = new Path("file:///D:/1.sh");
58              //从本地(Windows系统)上传文件到HDFS
59              fs.copyFromLocalFile(p2, p);
60              fs.close();
61          }
62      }
```

(3)测试代码。

① 在虚拟机 qf01 中,单独开启 HDFS 进程。

```
[root@qf01 ~]# start-dfs.sh
```

② 进行单元测试。单元测试是指对一个模块、一个函数或者一个类进行正确性检验的测试工作。例 3-1 的代码共有 3 个单元:将数据写入 HDFS 文件,读取 HDFS 文件,从 Windows 上传文件到 HDFS。以"将数据写入 HDFS 文件"单元为例,单元测试的步骤如下。

在 IntelliJ IDEA 中,单击程序 OperateHDFS.java 中的 "Run Test" 按钮。在出现的列表中,单击 "Run 'writeToHDFS()'" 选项运行程序。

③ 测试结果。运行方法 writeToHDFS()后,在虚拟机 qf01 的终端中输入以下命令。

```
[root@qf01 ~]# hdfs dfs -cat /1.txt
helloworld
```

输出 helloworld 即正确。

运行方法 readHDFSFile()后,控制台显示如下内容即正确。

```
C:\mysoft\Java\jdk1.8.0_121\bin\java...
log4j:WARN No appenders could be found for logger (org.apache.hadoop.util.Shell).
log4j:WARN Please initialize the log4j system properly.
log4j:WARN See http://logging.apache.org/log4j/1.2/faq.html#noconfig for more info.
helloworld
Process finished with exit code 0
```

程序运行结束后,在虚拟机 qf01 的终端中输入以下命令

```
[root@qf01 ~]# hdfs dfs -cat /1.sh
Hello Xiaoqian
```

输出 Hello Xiaoqian 即正确。

至此,使用 Java 程序操作 HDFS 完成。

3.5 Hadoop 序列化

3.5.1 Hadoop 序列化简介

1. 序列化和反序列化

序列化是将对象转化为字节流,以便在网络间进行传输或者在磁盘上永久存储的过程。反序列

化是将字节流转换为原对象的过程。序列化在 Hadoop 生态圈中有两个主要应用领域：进程间通信和永久存储。

2. Hadoop 序列化概述

（1）序列化机制在 Hadoop 中的应用。

Hadoop 实现进程间通信的 RPC（Remote Procedure Call，远程过程调用）中就使用了序列化机制。RPC 可以将消息序列转化成字节流后发送到远程节点，远程节点接着将字节流反序列化为原始消息。RPC 的序列化实现了紧凑、快速、可扩展、兼容性好的格式，能够使用户充分利用网络带宽这一稀缺资源。

（2）Hadoop 序列化的实现。

Hadoop 的序列化是通过 Writable 接口来实现的。Writable 接口的源码如下。

```
package org.apache.hadoop.io;
...
public interface Writable {
    void write(DataOutput var1) throws IOException;
    void readFields(DataInput var1) throws IOException;
}
```

由以上源码可知，Writable 接口包含两个方法：write(DataOutput out)与 readFields(DataInput in)。write()用来将数据写入指定的流，readFields()用来从指定的流中读取数据。

（3）Hadoop 不使用 Java 的序列化的原因。

Java 序列化机制的缺点：计算时开销大；序列化的结果占用存储空间大，序列化后的对象有时比序列化前的对象大数倍乃至十几倍；Java 序列化的引用机制导致大文件不能分割。这些缺点使得 Java 的序列化机制对 Hadoop 来说是不合适的。于是 Hadoop 实现了一套自己的序列化框架，相对于 Java 的序列化更简洁，集群信息传递的速度更快，所占存储空间更小。

3.5.2 常用实现 Writable 接口的类

由于实现 Writable 接口的类将 Java 的数据类型进行了序列化，因此可以把这些类看作 Java 数据类型的封装。Hadoop 实现 Writable 接口的类与 Java 数据类型的对应关系，如表 3.1 所示。

表 3.1   Hadoop 实现 Writable 接口的类与 Java 数据类型的对应关系

Hadoop 实现 Writable 接口的类	Java 数据类型
BooleanWritable	boolean
ByteWritable	byte
IntWritable	int/char
ShortWritable	short
LongWritable	long
FloatWritable	float
DoubleWritable	double
Text	String
ArrayWritable	Array

实现 Writable 接口的类大多在 Hadoop API 中的 org.apache.hadoop.io 包下，详见 Hadoop 官方网站。

在 Hadoop 实现 Writable 接口的类中，最常用的是 IntWritable 类和 Text 类，下面对这两个类进行简单讲解。

1. IntWritable 类

通过查看 Hadoop API 可知，IntWritable 类直接实现的是 Writable 接口的子接口 WritableComparable。WritableComparable 接口除了继承了 Writable 接口，还继承了 java.lang.Comparable 接口。WritableComparable 接口对于后续学习 MapReduce 非常重要。IntWritable 类的部分源码如下。

```java
public class IntWritable implements WritableComparable<IntWritable> {
    private int value;
    public IntWritable() {
    }
    public IntWritable(int value) {
        this.set(value);
    }
    public void set(int value) {
        this.value = value;
    }
    public int get() {
        return this.value;
    }
    public void readFields(DataInput in) throws IOException {
        this.value = in.readInt();
    }
    public void write(DataOutput out) throws IOException {
        out.writeInt(this.value);
    }
}
```

由以上源码可知，IntWritable 类在其内部封装了一个 int 类型的数据，通过 set() 与 get() 方法获取和设置其值，通过 readFields() 与 write() 方法对其进行输出与输入操作。

2. Text 类

Text 类和 Java 中的 String 类都使用标准的 unicode（统一码），两者的区别如下。

（1）两者编码方式不同。

Text 类使用的编码是 UTF-8，String 类使用的编码是 UTF-16。

（2）两者值的长度、索引位置的含义不同。

Text 值的索引位置是指 UTF-8 编码后的字节偏移量，Text 值的长度是指 UTF-8 编码后的字节数组大小；而 String 值的索引位置是指字符在字符串中的位置，String 值的长度是指字符的个数。

Text 类的大多数操作与 Java 的 String 类一样，但是当使用多字节的值时，就可以看出两者的区别。接下来编写程序，来验证 Text 类和 Java 的 String 类的区别。在 IntelliJ IDEA 中 testHDFS 模块的 src/main/java 文件夹下，新建类 com.qf.serialize.TestText。

【例 3-2】TestText.java

```
1   public class TestText {
2       @Test
3       public void string () throws UnsupportedEncodingException{
4           String s= "\u5c0f\u5343\u5c0f\u950b";   //使用多字节的值
5           System.out.println("String:"+s);
6           System.out.println(s.length());
7           System.out.println(s.indexOf("\u5343"));
8           System.out.println(s.indexOf("\u950b"));
9           System.out.println(s.charAt(0));
10      }
11      @Test
```

```
12      public void text(){
13          Text t=new Text("\u5c0f\u5343\u5c0f\u950b");
14          System.out.println("Text:"+t);
15          System.out.println(t.getLength());
16          System.out.println(t.find("\u5343"));
17          System.out.println(t.find("\u950b"));
18          System.out.println(t.charAt(0));
19      }
20  }
```

运行结果如下。

```
String:小千小锋
4
1
3
小
Text:小千小锋
12
3
9
23567
```

通过比对运行结果,可以直观地发现 Text 类与 Java 的 String 类的值的长度、索引位置存在区别。

3.5.3 自定义实现 Writable 接口的类

Hadoop 自带一系列实现 Writable 接口的类,绝大多数情况下可以满足用户的需求,但有时需要自定义实现 Writable 接口的类,来应对复杂的业务需求。

下面通过编程来自定义实现 Writable 接口的类,本书以自定义实现 Writable 的子接口 WritableComparable 为例。在 IntelliJ IDEA 中 testHDFS 模块的 src/main/java 文件夹下,新建类 com.qf. serialize.PersonWritable。

【例 3-3】PersonWritable.java

```
1   public class PersonWritable implements WritableComparable<PersonWritable> {
2       //定义 IntWritable 类型数据
3       private IntWritable age;
4       //定义 Text 类型数据
5       private Text name;
6       public PersonWritable() {
7       }
8       public PersonWritable(IntWritable age, Text name) {
9           this.age = age;
10          this.name = name;
11      }
12      public IntWritable getAge() {
13          return age;
14      }
15      public Text getName() {
16          return name;
17      }
18      public void setAge(IntWritable age) {
19          this.age = age;
20      }
```

```java
21      public void setName(Text name) {
22          this.name = name;
23      }
24      //比较方法，使得对象可比较大小
25      public int compareTo(PersonWritable o) {
26          int cmp = age.compareTo(o.getAge());
27          if (cmp != 0) return cmp;
28          return name.compareTo(o.getName());
29      }
30      @Override
31      //序列化方法
32      public void write(DataOutput out) throws IOException {
33          age.write(out);
34          name.write(out);
35      }
36      @Override
37      //反序列化方法
38      public void readFields(DataInput in) throws IOException {
39          age.readFields(in);
40          name.readFields(in);
41      }
42      @Override
43      public boolean equals(Object o) {
44          if (this == o) return true;
45          if (o == null || getClass() != o.getClass()) return false;
46          PersonWritable that = (PersonWritable) o;
47          return Objects.equals(age, that.age) &&
48                  Objects.equals(name, that.name);
49      }
50      @Override
51      public int hashCode() {
52          return Objects.hash(age, name);
53      }
54      @Override
55      public String toString() {
56          return "PersonWritable{" +
57                  "age=" + age +
58                  ", name=" + name +
59                  '}';
60      }
61  }
```

3.6 Hadoop 小文件处理

Hadoop 是为处理大型文件而设计的，在小文件的处理上效率较低，然而在实际生产环境中，需要 Hadoop 处理的数据往往存放在海量小文件中，因此，高效处理小文件对于提高 Hadoop 的性能至关重要。这里的小文件是指小于 HDFS 中一个块（Block）大小的文件。

Hadoop 处理小文件有两种方法：压缩小文件和创建序列文件。

3.6.1 压缩小文件

Hadoop 在存储海量小文件时，需要频繁访问各节点，非常耗费资源。如果某个节点上存放 1000

万个 600Byte 的文件，那么该节点至少需要提供 4 GB 的内存。为了节省资源，海量小文件在存储到 HDFS 之前，需要进行压缩。

1. Hadoop 压缩格式

Hadoop 进行文件压缩的作用：减少存储空间占用，降低网络负载。这两点对于 Hadoop 存储和传输海量数据非常重要。Hadoop 常用的压缩格式，如表 3.2 所示。

表 3.2　　　　　　　　　　　　　Hadoop 常用的压缩格式

压缩格式	是否可切分文件	算法	扩展名	Linux 压缩工具
Bzip2	是	Bzip2	.bz2	bzip2
DEFLATE	否	DEFLATE	.deflate	无
Gzip	否	DEFLATE	.gz	gzip
LZO	是（需要加索引）	LZO	.lzo	lzop
Snappy	否	Snappy	.snappy	无
LZ4	否	LZ4	.lz4	无

2. 编解码器

编解码器（Codec）是用于压缩和解压缩的设备或计算机程序。Hadoop 中的压缩编解码器主要是通过 Hadoop 的一些类来实现的，如表 3.3 所示。

表 3.3　　　　　　　　　　　　Hadoop 实现压缩编解码器的类

压缩格式	实现压缩编解码器的类
Bzip2	org.apache.hadoop.io.compress. BZip2Codec
DEFLATE	org.apache.hadoop.io.compress. DeflateCodec
Gzip	org.apache.hadoop.io.compress. GzipCodec
LZO	com.hadoop.compression.lzo.LzoCodec
Snappy	org.apache.hadoop.io.compress. SnappyCodec
LZ4	org.apache.hadoop.io.compress. Lz4Codec

对于 LZO 压缩格式，Hadoop 实现压缩编解码器的类不在 org.apache.hadoop.io.compress 包中，可前往 GitHub 官方网站下载。

3. 压缩格式的效率

对上述 6 种压缩格式的压缩效率、解压效率、压缩占比进行测试，测试结果如下。

压缩效率（由高到低）：Snappy > LZ4 > LZO > Gzip > DEFLATE > Bzip2
解压效率（由高到低）：Snappy > LZ4 > LZO > Gzip > DEFLATE > Bzip2
压缩占比（由小到大）：Bzip2 < DEFLATE < Gzip < LZ4 < LZO < Snappy

在实际生产环境中，可以参考以上的测试结果，根据业务需要做出恰当的选择。

3.6.2　创建序列文件

创建序列文件主要是指创建 SequenceFile（顺序文件）和 MapFile（映射文件）。

1. SequenceFile

（1）SequenceFile 简介。

SequenceFile 是存储二进制键值（Key-Value）对的持久数据结构。通过 SequenceFile 可以将若干小文件合并成一个大的文件进行序列化操作，实现文件的高效存储和处理。

（2）SequenceFile 的内部结构。

SequenceFile 由一个文件头（Header）和随后的一条或多条记录（Record）组成（见图 3.17）。

SequenceFile 中，Header 的前三个字节是 SEQ（顺序文件代码），随后的一个字节是 SequenceFile 的版本号。Header 还包括 Key 类的名称、Value 类的名称、压缩细节、Metadata（元数据）、Sync Marker（同步标识）等。Sync Marker 的作用在于可以读取 SequenceFile 任意位置的数据。

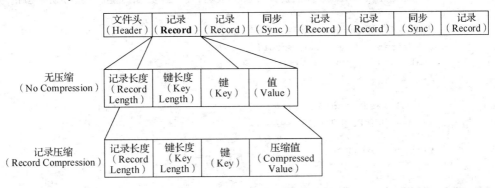

图 3.17 SequenceFile 的内部结构

记录有无压缩、记录压缩、块压缩三种压缩形式，默认为无压缩。

① 当采用无压缩（No Compress）时，每条记录由记录长度、键长度、键、值组成，将键与值序列化写入 SequenceFile。

② 当采用记录压缩（Record Compress）时，只压缩值，不压缩键，其他方面与无压缩类似。

③ 块压缩（Block Compress）利用记录间的相似性进行压缩，一次性压缩多条记录，比单条记录的压缩方法压缩效率更高。

当采用块压缩时，多条记录被压缩成默认 1MB 的数据块，每个数据块之前插入同步标识。数据块由表示数据块字节数的字段和压缩字段组成，其中，压缩字段包括键长度、键、值长度、值。

下面通过编程来理解 SequenceFile 的读写操作。

（3）SequenceFile 写操作。

① 在虚拟机 qf01 上启动 ZooKeeper 集群和 Hadoop 集群。查看处于活跃状态的 NameNode，以下操作需要在活跃的 NameNode 上进行。本书以 nn1 处于活跃状态为例进行讲述。

② 在虚拟机 qf01 上新建空文件 MySequenceFile.seq，并上传到 HDFS 的根目录下。

```
[root@qf01 ~]# vi MySequenceFile.seq
[root@qf01 ~]# hdfs dfs -put MySequenceFile.seq /
```

③ 编写程序。

【例 3-4】SequenceFileWriter.java

```
1    public class SequenceFileWriter {
2        /**
3         * 测试 SequenceFile 写操作
4         */
5        public static void main(String[] args) throws Exception {
6            //1.指定文件系统
7            Configuration conf = new Configuration();
8            conf.set("fs.defaultFS", "hdfs://192.168.142.131:9000");
9            //2.设置文件输出路径
10           Path outputPath = new Path("/MySequenceFile.seq");
11           //3.指定写操作的路径和内容
12           Option optPath = SequenceFile.Writer.file(outputPath);
```

```
13        Option optKey = SequenceFile.Writer.keyClass(IntWritable.class);
14        Option optVal = SequenceFile.Writer.valueClass(Text.class);
15        //4.执行写操作
16        SequenceFile.Writer writer = SequenceFile.createWriter(conf, optPath, optKey, optVal);
17        //5.向文件MySequenceFile.seq添加内容
18        IntWritable key = new IntWritable();
19        Text value = new Text();
20        for (int i = 0; i < 100; i++) {
21            key.set(i);
22            value.set("tom" + i);
23            writer.append(key, value);
24        }
25        writer.close();
26    }
27 }
```

(4) SequenceFile 读操作。

【例3-5】SequenceFileReader.java

```
1  public class SequenceFileReader {
2      /**
3       * 测试SequenceFile读操作
4       */
5      public static void main(String[] args) throws Exception {
6          //1.指定文件系统
7          Configuration conf = new Configuration();
8          conf.set("fs.defaultFS", "hdfs://192.168.142.131:9000");
9          //2.设置文件读取路径
10         Path path = new Path("/MySequenceFile.seq");
11         //3.指定读操作的路径
12         Option optPath = SequenceFile.Reader.file(path);
13         //4.执行读操作
14         SequenceFile.Reader reader = new SequenceFile.Reader(conf, optPath);
15         //5.将位置信息、写操作时写入的数据读取出来
16         Writable key = (Writable) ReflectionUtils.newInstance(reader.getKeyClass(), conf);
17         Writable value = (Writable) ReflectionUtils.newInstance(reader.getValueClass(), conf);
18         long position = reader.getPosition();
19         while (reader.next(key, value)) {
20             String syncSeen = reader.syncSeen() ? "*" : "";
21             System.out.printf("[%s%s]\t%s\t%s\n", position, syncSeen, key, value);
22             position = reader.getPosition();
23         }
24         reader.close();
25     }
26 }
```

(5) 分别执行例3-4和例3-5中的程序,在IntelliJ IDEA的控制台中出现以下内容表明SequenceFile的写操作和读操作均正确。

```
1    [128]    0    Tom0
2    [153]    1    Tom1
```

3	[178]	2	Tom2
4	[203]	3	Tom3
5	[228]	4	Tom4
6	[253]	5	Tom5
7	[278]	6	Tom6
8	[303]	7	Tom7
9	[328]	8	Tom8
10	[353]	9	Tom9
11	[378]	10	Tom10
12	[404]	11	Tom11
13	[430]	12	Tom12
14	[456]	13	Tom13
15	[482]	14	Tom14

2. MapFile

MapFile 可以看作已排序的 SequenceFile，是一个可供查询的文件类。MapFile 内部维护了两个 SequenceFile 目录：Index 目录和 Data 目录。Index 目录存放的是索引文件（包含键和偏移量），能够加载到内存，主要用于对数据文件的快速查找。Data 目录存放的是数据文件，记录了通过键查找的值。MapFile 的写操作和读操作与 SequenceFile 相似，参考 HadoopAPI 中的 org.apache.hadoop.io.MapFile 类进行测试即可。

3.7 通信机制 RPC

3.7.1 RPC 简介

1. RPC 的概念

Hadoop 中节点的进程间通信是通过 RPC 来实现的。RPC（Remote Procedure Call，远程过程调用）是一种通过网络从远程计算机程序上请求服务的协议。在使用 RPC 时，用户不需要了解底层网络技术。RPC 采用客户机/服务器模式。RPC 客户机指发出请求的程序，RPC 服务器指提供服务的程序。

RPC 具有封装网络交互、支持容器、可配置、可扩展等特点。RPC 使得开发含有网络分布式程序的应用程序更加容易。

2. 远程过程调用的执行步骤

（1）在客户机端，调用进程发送一个有进程参数的调用信息到服务进程，然后等待应答信息。

（2）在服务器端，服务进程保持睡眠状态，直到调用信息到达。

（3）当调用信息到达时，服务器获得进程参数，计算出结果，发送应答信息，等待下一个调用信息。

（4）客户机调用进程接收应答信息，获得进程结果，然后循环执行下去。

3.7.2 Hadoop 的 RPC 架构

Hadoop 进程间通信的 RPC 架构包含 RPC Server、RPC Client、RPC Interface 三个节点，如图 3.18 所示。RPC Server 负责服务器端的实现，RPC Client 负责客户机端的实现，RPC Interface 负责对外接口类的实现。

使用 Hadoop RPC 有以下几个步骤。

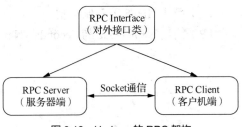

图 3.18 Hadoop 的 RPC 架构

（1）定义 RPC 服务接口。
RPC 定义了服务器端对外提供的服务接口。
（2）实现 RPC 服务接口。
Hadoop 的 RPC 服务接口通常是一个 Java 接口，用户需要实现该接口。
（3）构造和启动 RPC Server。
使用 Builder 类构造 RPC Server，并调用 start()方法启动 RPC Server。
（4）RPC Client 发送请求给 RPC Server。
RPC Client 调用线程发起 RPC 连接请求，等待 RPC Server 响应后，向其传输数据。

3.8 本章小结

本章主要讲解 Hadoop 的分布式文件系统。首先讲解了 HDFS 的基本概念和特点，以及 HDFS 的架构和原理，让大家对 HDFS 内部运行有一定的理解；然后对 HDFS 操作进行了讲解，并编写程序进行实践；最后讲解了 Hadoop 的序列化和 Hadoop 对小文件的处理方式，让大家可以更好地理解 Hadoop 的内部原理，以及优化方法。通过对本章内容的学习，应达到熟练使用 HDFS 分布式文件系统来管理文件的目标。

3.9 习题

1. 填空题

（1）下载 HDFS 文件或目录到本地的命令是＿＿＿＿＿。
（2）HDFS 集群节点包括＿＿＿＿＿、DataNode、SecondaryNameNode。
（3）将对象转化为字节流，以便在网络间进行传输或者在磁盘上永久存储的过程是＿＿＿＿＿。
（4）RPC 具有的特点：＿＿＿＿＿、＿＿＿＿＿、＿＿＿＿＿、＿＿＿＿＿。
（5）NameNode 与 DataNode 的进程间通信是通过＿＿＿＿＿实现的。

2. 选择题

（1）HDFS 中的 Block 默认保存（   ）。
    A. 3 份　　　　　　B. 2 份　　　　　　C. 1 份　　　　　　D. 不确定
（2）Hadoop 的序列化是通过（   ）接口来实现的。
    A. Text　　　　　　B. IntWritable　　　C. Writable　　　　D. ByteWritable
（3）序列化在 Hadoop 生态圈中的两个主要应用领域是（   ）和（   ）。
    A. 压缩文件　　　　B. 永久存储　　　　C. 解压文件　　　　D. 进程间通信

3. 思考题

（1）简述 HDFS 的存储机制。
（2）简述 NameNode 与 SecondaryNameNode 的区别与联系。

第4章 MapReduce分布式计算框架

本章学习目标
- 理解 MapReduce 的基本原理
- 理解 MapReduce 经典案例 WordCount 的实现原理
- 掌握 MapReduce 运行流程
- 掌握 MapReduce 程序设计方法

Hadoop 的数据处理核心为 MapReduce 分布式计算框架。这一框架的出现，使得编程人员在不熟悉分布式并行编程的情况下，可以将自己的程序运行在分布式系统上来处理海量的数据，因此，大数据开发人员需要重点掌握 MapReduce 的基本原理。

4.1 认识 MapReduce

4.1.1 MapReduce 核心思想

MapReduce 的核心思想是将大数据分而治之，即将数据通过一定的数据划分方法，分成多个较小的具有同样计算过程的数据块，数据块之间不存在依赖关系，将每一个数据块分给不同的节点去处理，最后将处理的结果汇总。

具体来说，对大量顺序式数据元素或者记录进行扫描和对每个数据元素或记录做相应的处理并获得中间结果的两个过程被抽象为 Map 操作，对中间结果进行收集整理和产生最终结果并输出的过程被抽象为 Reduce 操作。

MapReduce 提供统一框架来隐藏系统层的细节，实现了自动并行处理，如计算任务的自动划分和调度、数据的自动化分布式存储和划分、处理数据与计算任务的同步、结果数据的收集整理、系统通信、负载平衡、计算性能优化处理、处理节点出错检测和失效恢复等。

4.1.2 MapReduce 编程模型

MapReduce 是一种分布式离线并行计算框架，主要用于大规模数据集（大于 1TB）的并行计算。Hadoop MapReduce 可以看作 Google MapReduce 的克隆版。

MapReduce 的特点是易于编程，具有良好的扩展性，具有高容错性，适合 PB 级以上海量数据的离线处理。MapReduce 的两大核心思想是 Map（映射）和 Reduce（化简）。基于这两大核心思想，MapReduce 把数据处理流程分成两个主要阶段：Map 阶段和 Reduce 阶段。

Map 阶段负责对数据进行预处理，具体是指通过特定的输入格式读取文件数据，将读取的数据以键值（Key-Value，K-V）对的形式进行保存。

Reduce 阶段负责对数据进行聚合处理，具体是指通过对 Map 阶段保存的数据进行归并、排序等，计算出想要的结果。

MapReduce 的整体结构如图 4.1 所示。

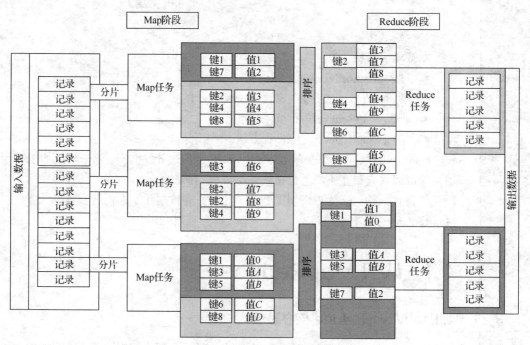

图 4.1　MapReduce 结构图

从结构图中可以看出 MapReduce 的处理过程。

（1）Map 阶段对数据进行分块和分片处理。

（2）将数据信息交给 Map 任务去进行读取。

（3）对数据进行分类后写入，根据不同的键产生相应的键值对数据。

（4）进入 Reduce 阶段，执行定义的算法，使用相同键的值从多个数据表中被集合到一起进行分类处理，最终结果输出到相应的磁盘空间。

MapReduce 的框架流程如图 4.2 所示。

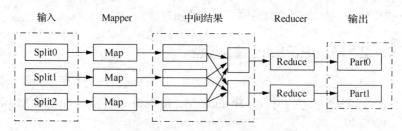

图 4.2　MapReduce 的框架流程

从图 4.2 中看到，Hadoop 为每个创建的 Map 任务分配输入文件的一部分，这部分称为 Split（切片）。用户自定义的 Map 能够根据用户需要来处理每个 Split 中的内容。

Map 读取 Split 内容后，将其解析成键值对的形式进行运算，并将 Map 中定义的算法应用至每一条内容，而内容范围可以根据用户自定义来确定。

一个 Reduce 任务的执行分成三个阶段。

（1）Reduce 获取 Map 输入的处理结果。

（2）对拥有相同键值对的数据进行分组。

（3）将用户定义的算法应用到每个键值对的数据列表。

在实际开发环境中，基于 MapReduce 的计算模型大大降低了分布式并行编程的难度。程序员只需编写 map() 和 reduce() 两个方法，就可以完成简单的大数据处理任务，而并行编程中的其他复杂问题，如作业调度、负载均衡、容错处理、网络通信等，均由 YARN 负责处理。

下面通过 MapReduce 实现经典应用 WordCount（单词统计），来更直观地了解 MapReduce 的使用方法。

4.1.3　MapReduce 编程案例——WordCount

1. WordCount 问题梳理

问题：现有一批英文文件，规模为 PB 级，如何统计这些文件中某单词出现的次数？

MapReduce 的解决方案如下。

（1）分别统计每个文件中该单词出现次数，即编写 map() 方法。

（2）累加不同文件中该单词出现次数，即编写 reduce() 方法。

2. MapReduce 编程的一般思路

（1）输入一系列键值对(K1,V1)。

（2）通过 map() 方法和 reduce() 方法处理输入的键值对。

① 用 map() 方法将(K1,V1)处理成 list(K2,V2) 的形式。

② 用 reduce() 方法将(K2,list(V2))处理成 list(K3,V3) 的形式。

简略表示如下。

Map 阶段：(K1, V1)→list(K2, V2)。

Reduce 阶段：(K2, list(V2))→list(K3, V3)。

（3）输出一系列键值对(K3,V3)。

3. MapReduce 编程模型

MapReduce 编程模型是 MapReduce 编程的基本格式，其他更加复杂的编程也基于这一模型。MapReduce 编程模型主要分为以下三部分，程序也是按照这个顺序来编写。

（1）实现 Mapper。代码如下。

```
public class MyMapper extends Mapper ... {
    //重写 map()方法
}
```

（2）实现 Reducer。代码如下。

```
public class MyReducer extends Reducer ... {
    //重写 reduce()方法
}
```

（3）创建 MapReduce 作业。MapReduce 作业即 MapReduce 应用程序（Application）。创建 MapReduce 作业的核心代码如下。

```java
public class MyApp {
    public static void main(String[] args) {
        //1.新建配置对象，为配置对象设置文件系统
        Configuration conf = new Configuration();
        conf.set("fs.defaultFS", "hdfs://192.168.142.131:9000");
        //2.设置Job属性
        Job job = Job.getInstance(conf);   //通过 Configuration 获得 Job 实例
        job.setJobName("MyApp");    //为 Job 命名
        job.setJarByClass(MyApp.class);    //为 Job 运行设置主类
        //3.设置数据输入路径
        Path inPath = new Path(args[0]);
        FileInputFormat.addInputPath(job, inPath);
        //4.设置 Job 执行的 Mapper 类和输出 K-V 类型
        job.setMapperClass(MyMapper.class);
        job.setMapOutputKeyClass(Text.class);
        job.setMapOutputValueClass(IntWritable.class);
        //5.设置 Job 执行的 Reducer 类和输出 K-V 类型
        job.setReducerClass(MyReducer.class);
        job.setOutputKeyClass(Text.class);
        job.setOutputValueClass(IntWritable.class);
        //6.设置数据输出路径
        Path outPath = new Path(args[1]);
        FileOutputFormat.setOutputPath(job, outPath);
        //7.MapReduce 作业完成后退出系统
        System.exit(job.waitForCompletion(true) ? 0 : 1);
    }
}
```

4. 配置 MapReduce 开发环境

在开始编写程序之前，仍然需要配置 MapReduce 开发环境。在 IntelliJ IDEA 中新建模块 testMapReduce，修改 pom.xml 文件内容如下。

```xml
<?xml version="1.0" encoding="UTF-8"?>
<project xmlns="http://maven.apache.org/POM/4.0.0"
     xmlns:xsi="http://www.w3.org/2001/XMLSchema-instance"
     xsi:schemaLocation="http://maven.apache.org/POM/4.0.0 http://maven.apache.org/xsd/maven-4.0.0.xsd">
    <modelVersion>4.0.0</modelVersion>
    <groupId>com.qf</groupId>
    <artifactId>testMapReduce</artifactId>
    <version>1.0-SNAPSHOT</version>
    <dependencies>
        <!--Hadoop 客户端依赖，该依赖包含 HDFS 的相关依赖-->
        <dependency>
            <groupId>org.apache.hadoop</groupId>
            <artifactId>hadoop-client</artifactId>
            <version>2.7.3</version>
        </dependency>
        <!--单元测试的依赖-->
```

```
            <dependency>
                <groupId>junit</groupId>
                <artifactId>junit</artifactId>
                <version>4.12</version>
            </dependency>
        </dependencies>
</project>
```

5. 实现 Mapper

注意：org.apache.hadoop.mapreduce 是 Hadoop 新 API 中的包，本书使用的是这个包下的 API，而 org.apache.hadoop.mapred 是 Hadoop 旧 API 中的包，在导入时不要弄错。

【例 4-1】MyMapper.java

```
1   public class MyMapper extends Mapper<LongWritable, Text, Text, IntWritable> {
2       Text word = new Text();
3       IntWritable one = new IntWritable(1);
4       @Override
5       protected void map(LongWritable key, Text value, Context context)
6               throws IOException, InterruptedException {
7           //1.以行为单位，对数据进行处理
8           String line = value.toString();
9           //2.以空格为分隔符，对单词进行拆分
10          String[] words = line.split(" ");
11          //3.迭代数组，将输出的 K-V 对存入 context
12          for (String s : words) {
13              word.set(s);
14              context.write(word, one);
15          }
16      }
17  }
```

6. 实现 Reducer

【例 4-2】MyReducer.java

```
1   public class MyReducer extends Reducer<Text, IntWritable, Text, IntWritable> {
2       @Override
3       protected void reduce(Text key, Iterable<IntWritable> values, Context context) throws IOException, InterruptedException {
4           //1.定义一个计数器
5           Integer counter = 0;
6           //2.迭代数组，将输出的 K-V 对存入 context
7           for (IntWritable value : values) {
8               counter += value.get();
9           }
10          context.write(key, new IntWritable(counter));
11      }
12  }
```

7. 创建 MapReduce 作业

【例 4-3】WordCountApp.java

```
1   public class WordCountApp{
2       public static void main(String[] args) throws Exception {
3           if (args == null || args.length < 2) {
```

```
4           throw new Exception("参数不足,需要两个参数!");
5       }
6       //1.新建配置对象,为配置对象设置文件系统
7       Configuration conf = new Configuration();
8       conf.set("fs.defaultFS", "hdfs://192.168.142.131:9000");
9       //2.设置 Job 属性
10      Job job = Job.getInstance(conf, "WordCountApp");
11      job.setJarByClass(WordCountApp.class);
12      //3.设置数据输入路径
13      Path inPath = new Path(args[0]);
14      FileInputFormat.addInputPath(job, inPath);
15      //4.设置 Job 执行的 Mapper 类和输出 K-V 类型
16      job.setMapperClass(MyMapper.class);
17      job.setMapOutputKeyClass(Text.class);
18      job.setMapOutputValueClass(IntWritable.class);
19      //5.设置 Job 执行的 Reducer 类和输出 K-V 类型
20      job.setReducerClass(MyReducer.class);
21      job.setOutputKeyClass(Text.class);
22      job.setOutputValueClass(IntWritable.class);
23      //6.设置数据输出路径
24      Path outPath = new Path(args[1]);
25      FileOutputFormat.setOutputPath(job, outPath);
26      //7.MapReduce 作业完成后退出系统
27      System.exit(job.waitForCompletion(true) ? 0 : 1);
28     }
29 }
```

8. 在 Hadoop 集群上运行 MapReduce 作业

(1) 将程序打包成 jar 包。

在 IntelliJ IDEA 中,单击右侧边栏的"Maven Projects"栏,在出现的下拉列表中,依次双击"testMapReduce""Lifecycle""package"选项,在控制台出现"BUILD SUCCESS",表明已将程序打包成 jar 包。

(2) 在虚拟机 qf01 上的/root 目录(root 用户的~目录)下,创建文件 word.txt,并输入以下内容。

```
hello qianfeng hi qianfeng
hello world
hi hadoop hi mapreduce
```

(3) 本书使用的是 HA 模式下的 Hadoop 集群,需要启动 ZooKeeper 集群和 Hadoop 集群。

```
[root@qf01 ~]# zkServer.sh start
[root@qf01 ~]# start-dfs.sh
[root@qf01 ~]# start-yarn.sh
```

(4) 查看处于活跃状态的虚拟机。本书都以虚拟机 qf01 处于活跃状态为例进行讲述。在虚拟机 qf01 上,上传 word.txt 文件到 HDFS 的根目录下。

```
[root@qf01 ~]# hdfs dfs -put word.txt /
```

(5) 将 testMapReduce-1.0-SNAPSHOT.jar 发送到虚拟机 qf01。

在左侧边栏的"Project"栏中,依次双击"testHadoop""testMapReduce""target",在出现的 testMapReduce-1.0-SNAPSHOT.jar 上单击鼠标右键,在出现的列表中单击"Show in Explorer"选项,

将弹出的文件管理器中的 testMapReduce-1.0-SNAPSHOT.jar 发送到虚拟机 qf01 的 /root 目录下。

（6）在虚拟机 qf01 上，运行 MapReduce 作业的常用格式如下。

```
hadoop jar Jar包 完整包名类名 待处理文件在 HDFS 上的绝对路径 文件处理后在 HDFS 上的存放目录
```

运行 MapReduce 作业。

```
[root@qf01 ~]# hadoop jar testMapReduce-1.0-SNAPSHOT.jar com.qf.mr.wordcount.
WordCountApp /word.txt /outdata
```

（7）查看运行结果有两种方法。

① 在虚拟机 qf01 上，使用如下命令查看运行结果。

```
[root@qf01 ~]# hdfs dfs -cat /outdata/part-r-00000
hadoop       1
hello        2
hi           3
mapreduce    1
qianfeng     2
```

其中，part-r-00000 是 MapReduce 作业运行后产生的文件。

② 登录 NameNode 的 Web 界面（http://192.168.142.131:50070），单击"Utilities"菜单下的"Browse the file system"选项，在搜索框中输入"/"，单击"Go!"按钮，出现 HDFS 根目录下的文件和目录，如图 4.3 所示。

图 4.3　HDFS 根目录下的文件和目录

依次单击"outdata""part-r-00000"，出现文件下载界面，如图 4.4 所示。单击"Download"按钮下载文件 part-r-00000，用记事本打开文件 part-r-00000，看到的结果与使用命令查看的结果一致。

图 4.4　文件下载界面

4.2 MapReduce 编程组件

4.2.1 InputFormat 组件

InputFormat（输入格式）组件用来对输入数据进行格式化。InputFormat 描述了 MapReduce 作业的输入规范。

1. InputFormat 源码

```
public abstract class InputFormat<K, V> {
  public abstract
    List<InputSplit> getSplits(JobContext context
                   ) throws IOException, InterruptedException;
  public abstract
    RecordReader<K,V> createRecordReader(InputSplit split,
                       TaskAttemptContext context
                     ) throws IOException,
                              InterruptedException;
}
```

由 InputFormat 源码可知，InputFormat 有两个方法：getSplits()和 createRecordReader()。

（1）getSplits()负责从逻辑上分割输入文件，得到输入切片（InputSplit），在获得 InputSplit 之前，需要考虑输入文件是否在逻辑上可分割、文件存储时分块的大小和文件大小等因素。InputSplit 在逻辑上包含了需要用单个 Mapper 处理的数据（键值对）。

（2）createRecordReader()负责创建 RecordReader（记录阅读器）。

2. InputFormat 的常用子类

InputFormat 的常用子类如表 4.1 所示。

表 4.1　　　　　　　　　　　　　　　　InputFormat 的常用子类

类名	继承自	处理的数据类型	特点
FileInputFormat	InputFormat	文本数据	InputFormat 的实现基类
TextInputFormat	FileInputFormat	文本数据	InputFormat 默认使用的子类，以行为单位处理数据
KeyValueTextInputFormat	FileInputFormat	文本数据	适合处理以下形式的数据：输入数据的每一行有两列，并用制表符分割
NLineInputFormat	FileInputFormat	文本数据	以 n 行为单位处理数据，切片按照指定的行号进行切分。在处理数据量小的情况下可以使用，减少分发过程
SequenceFileInputFormat	FileInputFormat	二进制数据	处理序列化的二进制数据
SequenceFileAsBinaryInputFormat	SequenceFileInputFormat	二进制数据	将 SequenceFile 的 Key 和 Value 转换为 Text 对象
SequenceFileAsTextInputFormat	SequenceFileInputFormat	二进制数据	将 SequenceFile 的 Key 和 Value 转换为二进制对象
FixedLengthInputFormat	FileInputFormat	二进制数据	从文件中读取固定长度的二进制记录
MultipleInputs	无	多种格式	处理多种格式的数据
DBInputFormat	InputFormat	关系型数据库（MySQL、Oracle）中的表	从关系型数据库中读取数据

FileInputFormat 主要用来设置 MapReduce 作业中数据的输入路径，使用的方法是 addInputPath()，WordCount 案例中就使用了这一方法。

```
FileInputFormat.addInputPath(job, inPath);
```

4.2.2 OutputFormat 组件

OutputFormat（输出格式）组件用来对输出数据进行格式化，可以将 MapReduce 作业的输出结果转换为其他系统可读的格式，从而实现与其他系统的互操作。OutputFormat 描述了 MapReduce 作业的输出规范。

在默认情况下，MapReduce 只有一个 Reduce，默认输出一个名为 part-r-00000 的文件，输出文件的个数与 Reduce 的个数一致。如果 MapReduce 有两个 Reduce，输出结果就有两个文件，第一个为 part-r-00000，第二个为 part-r-00001。如果 MapReduce 有更多 Reduce，以此类推。

1. OutputFormat 的主要源码

```
public abstract class OutputFormat<K, V> {
  public abstract RecordWriter<K, V>
    getRecordWriter(TaskAttemptContext context
                ) throws IOException, InterruptedException;
}
```

OutputFormat 源码中的 RecordWriter() 方法主要负责创建 RecordWriter（记录写入器）。

2. OutputFormat 的常用子类

OutputFormat 的常用子类如表 4.2 所示。

表 4.2　　　　　　　　　　　OutputFormat 的常用子类

类名	继承自	特点
FileOutputFormat	OutputFormat	OutputFormat 的实现基类
TextOutputFormat	FileOutputFormat	OutputFormat 默认使用的子类，把每条记录输出为文本行
NullOutputFormat	OutputFormat	什么都不输出
SequenceFileOutputFormat	FileOutputFormat	输出 SequenceFile
SequenceFileAsBinaryOutputFormat	SequenceFileOutputFormat	输出二进制 K-V 对格式的 SequenceFile
MapFileOutputFormat	FileOutputFormat	输出 MapFile
MultipleOutputs	无	将数据写到多个文件
DBOutputFormat	OutputFormat	将作业输出数据转存到关系型数据库中

3. 自定义 InputFormat 和 OutputFormat

以 InputFormat 的 SequenceFileInputFormat 和 OutputFormat 的 SequenceFileOutputFormat 为例，自定义 InputFormat 和 OutputFormat。其中，SequenceFileInputFormat 在 MapReduce 作业中应用的基本思路是在 HDFS 上创建 SequenceFile 源文件，实现 Mapper、Reducer，创建 MapReduce 作业。

（1）创建 SequenceFile 源文件。

【例 4-4】CreateSequenceFile.java

```
1    public class CreateSequenceFile {
2        public static void main(String[] args) throws Exception {
3            //1.指定文件系统
4            Configuration conf = new Configuration();
5            conf.set("fs.defaultFS", "hdfs://192.168.142.131:9000");
```

```
6       //2.设置文件输出路径
7       Path outputPath = new Path("/MySequenceFile2.seq");
8       //3.指定写操作的路径和内容
9       SequenceFile.Writer.Option optPath = SequenceFile.Writer.file(outputPath);
10      SequenceFile.Writer.Option optKey = SequenceFile.Writer.keyClass(Text.class);
11      SequenceFile.Writer.Option optVal = SequenceFile.Writer.valueClass(IntWritable.class);
12      //4.执行写操作
13      SequenceFile.Writer writer = SequenceFile.createWriter(conf, optPath, optKey, optVal);
14      //5.向文件 MySequenceFile.seq 中添加内容
15      writer.append(new Text("hello"), new IntWritable(10));
16      writer.append(new Text("qianfeng"), new IntWritable(20));
17      writer.append(new Text("hi"), new IntWritable(15));
18      writer.append(new Text("qianfeng"), new IntWritable(80));
19      writer.append(new Text("hello"), new IntWritable(6));
20      writer.append(new Text("world"), new IntWritable(7));
21      writer.close();
22    }
23  }
```

（2）实现 Mapper。

【例 4-5】MyMapper.java

```
1  public class MyMapper extends Mapper<Text, IntWritable, Text, IntWritable> {
2      @Override
3      protected void map(Text key, IntWritable value, Context context)
4          throws IOException, InterruptedException {
5          context.write(key, value);
6      }
7  }
```

（3）实现 Reducer。

【例 4-6】MyReducer.java

```
1  public class MyReducer extends Reducer<Text, IntWritable, Text, IntWritable> {
2      @Override
3      protected void reduce(Text key, Iterable<IntWritable> values, Context context) throws IOException, InterruptedException {
4          //1.定义一个计数器
5          Integer counter = 0;
6          //2.迭代数组,将输出的 K-V 对存入 context
7          for (IntWritable value : values) {
8              counter += value.get();
9          }
10         context.write(key, new IntWritable(counter));
11     }
12 }
```

（4）创建 MapReduce 作业。

【例 4-7】SequenceFileTestApp.java

```
1  public class SequenceFileTestApp {
2      public static void main(String[] args) throws Exception {
```

```
3            if (args == null || args.length < 2) {
4                throw new Exception("参数不足,需要两个参数!");
5            }
6            //1.新建配置对象,为配置对象设置文件系统
7            Configuration conf = new Configuration();
8            conf.set("fs.defaultFS", "hdfs://192.168.142.131:9000");
9            //2.设置 Job 属性
10           Job job = Job.getInstance(conf, "SequenceFileTestApp");
11           job.setJarByClass(SequenceFileTestApp.class);
12           //3.设置数据输入路径
13           Path inPath = new Path(args[0]);
14           FileInputFormat.addInputPath(job, inPath);
15           //4.设置 Job 的输入格式
16           job.setInputFormatClass(SequenceFileInputFormat.class);
17           //5.设置 Job 执行的 Mapper 类和输出 K-V 类型
18           job.setMapperClass(MyMapper.class);
19           job.setMapOutputKeyClass(Text.class);
20           job.setMapOutputValueClass(IntWritable.class);
21           //6.设置 Job 执行的 Reducer 类和输出 K-V 类型
22           job.setReducerClass(MyReducer.class);
23           job.setOutputKeyClass(Text.class);
24           job.setOutputValueClass(IntWritable.class);
25           //7.设置数据输出格式
26           //job.setOutputFormatClass(SequenceFileOutputFormat.class);
27           SequenceFileOutputFormat.setCompressOutput(job, true);
28           SequenceFileOutputFormat.setOutputCompressorClass(job,
GzipCodec.class);//设置压缩编解码器
29           SequenceFileOutputFormat.setOutputCompressionType(job,
SequenceFile.CompressionType.BLOCK);//设置压缩类型
30           //8.设置数据输出路径
31           Path outPath = new Path(args[1]);
32           FileOutputFormat.setOutputPath(job, outPath);
33           //9.MapReduce 作业完成后退出系统
34           System.exit(job.waitForCompletion(true) ? 0 : 1);
35       }
36   }
```

（5）在 HDFS 上创建 SequenceFile 源文件。

开启 Hadoop 集群的 HA 模式，在 IntelliJ IDEA 中运行 testMapReduce 模块的 com.qf.mr.inputformatandoutputformat 包下的 CreateSequenceFile 类。

（6）在 Hadoop 集群上运行 MapReduce 作业。

用 Maven Projects 将模块 testMapReduce 打成 Jar 包，将 testMapReduce-1.0-SNAPSHOT.jar 发送到活跃虚拟机的/root 目录下，运行 MapReduce 作业。

```
[root@qf01 ~]# hadoop jar testMapReduce-1.0-SNAPSHOT.jar com.qf.mr.
inputformatandoutputformat.SequenceFileTestApp /MySequenceFile2.seq /outdata/
inputformatandoutputformat
```

（7）通过 Web 界面下载 HDFS 中/outdata/inputformatandoutputformat 目录下的 part-r-00000.gz 压缩包，用记事本查看压缩包中的 part-r-00000 文件，得到 MapReduce 作业的运行结果如下。

hello   16

```
hi            15
qianfeng      100
world         7
```

以上自定义 InputFormat 为 SequenceFileInputFormat，未设置 OutputFormat，默认使用 TextOutputFormat。自定义 OutputFormat 为 SequenceFileOutputFormat 不再赘述。

4.2.3 RecordReader 组件和 RecordWriter 组件

1. RecordReader 组件

（1）InputSplit。

在学习 RecordReader 组件之前，需要先了解 InputSplit。

InputSplit 表示由单个 Mapper 处理的数据。通常，InputSplit 提供面向字节的输入视图，RecordReader 负责处理和呈现面向记录的视图。

（2）RecordReader。

RecordReader（记录阅读器）负责从每个 InputSplit 中读取键值对，以便输入到 Mapper 中，可以看作 InputSplit 的迭代器。RecordReader 的主要方法是 nextKeyvalue()，用来获取 Split 上的下一个键值对。

RecordReader 是一个抽象类，在 WordCount 案例中使用的是 RecordReader 的实现子类 LineRecordReader（行级记录阅读器）。LineRecordReader 以偏移量作为 Key（键），一行数据作为 Value（值）。

2. RecordWriter 组件

RecordWriter（记录写入器）的源码如下。

```
public abstract class RecordWriter<K, V> {
    public abstract void write(K key, V value) throws IOException, InterruptedException;
    public abstract void close(TaskAttemptContext context) throws IOException, InterruptedException;
}
```

由源码可知，RecordWriter 主要有两个方法：write()和close()。

（1）write()负责将 Reduce 输出的键值对写成特定的输出格式，以便输出到文件系统等。

（2）close()负责对输出做最后确认并关闭输出。

4.2.4 Partitioner 组件

Partitioner（分区）组件决定 Map 端输出的 Key 交由哪一个 Reduce 任务来处理。Partitioner 的数量与 Reduce 任务数相同，当 Reduce 任务数大于 1 时，设置 Partitioner 数量有效；当 Reduce 任务数为 1 时，设置 Partitioner 数量无效。Partitioner 的源码如下。

```
public abstract class Partitioner<KEY, VALUE> {
    public abstract int getPartition(KEY key, VALUE value, int numPartitions);
}
```

MapReduce 中默认使用的是 HashPartitioner 类，该类继承自 Partitioner 类。HashPartitioner 的源码如下。

```
public class HashPartitioner<K, V> extends Partitioner<K, V> {
    public int getPartition(K key, V value,int numReduceTasks) {
        return (key.hashCode() & Integer.MAX_VALUE) % numReduceTasks;
```

          }
      }

　　自定义 Partitioner 是解决常见的数据倾斜问题的方法之一。下面编写程序自定义 Partitioner 并进行测试，具体步骤如下。

　　（1）在 IntelliJ IDEA 中 testMapReduce 模块下新建包 com.qf.mr.partitioner。

　　（2）实现 Mapper 和 Reducer。使用 com.qf.mr.wordcount 包下的 MyMapper 和 MyReducer 即可，不需要再写程序。

　　（3）自定义 Partitioner。

【例 4-8】 TestPartitioner.java

```
1   public class TestPartitioner extends Partitioner<Text, IntWritable> {
2       /**
3        * 以 h 开头的单词放到 0 号分区，其他单词放到 1 号分区
4        */
5       @Override
6       public int getPartition(Text text, IntWritable intWritable, int numPartitions) {
7           String key = text.toString();
8           if (key.startsWith("h")) {
9               return 0;
10          } else {
11              return 1;
12          }
13      }
14  }
```

　　（4）创建 MapReduce 作业。

【例 4-9】 PartitionerApp.java

```
1   public class PartitionerApp {
2       public static void main(String[] args) throws Exception {
3           if (args == null || args.length < 2) {
4               throw new Exception("参数不足,需要两个参数!");
5           }
6           Configuration conf = new Configuration();
7           conf.set("fs.defaultFS", "hdfs://192.168.142.131:9000");
8           Job job = Job.getInstance(conf, "WordCountApp");
9           job.setJarByClass(PartitionerApp.class);
10          Path inPath = new Path(args[0]);
11          FileInputFormat.addInputPath(job, inPath);
12          job.setMapperClass(MyMapper.class);
13          job.setMapOutputKeyClass(Text.class);
14          job.setMapOutputValueClass(IntWritable.class);
15          //设置分区
16          job.setPartitionerClass(TestPartitioner.class);
17          job.setReducerClass(MyReducer.class);
18          job.setOutputKeyClass(Text.class);
19          job.setOutputValueClass(IntWritable.class);
20          //设置 reduce 任务的个数
21          job.setNumReduceTasks(2);
22          Path outPath = new Path(args[1]);
23          FileOutputFormat.setOutputPath(job, outPath);
```

```
24              System.exit(job.waitForCompletion(true) ? 0 : 1);
25          }
26      }
```

(5) 在 Hadoop 集群上运行 MapReduce 作业。

用 Maven Projects 将模块 testMapReduce 打成 Jar 包，将 testMapReduce-1.0-SNAPSHOT.jar 发送到活跃虚拟机的/root 目录下，运行 MapReduce 作业。

```
[root@qf01 ~]# hadoop jar testMapReduce-1.0-SNAPSHOT.jar com.qf.mr.partitioner.
PartitionerApp /word.txt /outdata/partitioner
```

(6) 在活跃的 NameNode 上查看 MapReduce 作业的运行结果。

登录 http://192.168.142.131:50070/explorer.html#/outdata/partitioner，查看 HDFS 的/outdata/partitioner 目录，如图 4.5 所示。

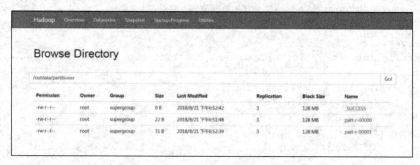

图 4.5　HDFS 的/outdata/partitioner 目录

图 4.5 中，在 HDFS 的/outdata/partitioner 目录下，出现两个文件 part-r-00000 和 part-r-00001。下载这两个文件到本地系统，用记事本查看文件内容。

① part-r-00000 文件的内容如下。

```
hadoop    1
hello     2
hi        3
```

② part-r-00001 文件的内容如下。

```
mapreduce    1
qianfeng     2
world        1
```

以 h 开头的单词在 part-r-00000 文件中，其他单词在 part-r-00001 文件中，这与例 4-8 中自定义 Partitioner 的要求相符，表明自定义 Partitioner 成功。

4.2.5　Combiner 组件

Combiner 组件用来对 Map 任务的输出数据进行合并，可以使输出数据更紧凑地写入磁盘或传送到 Reduce 阶段。Combiner 可以看作 Map 任务中的 Mini Reduce 任务。如果 Map 阶段设定了 Combiner，会在 Sort（排序）输出的基础上执行操作，Combiner 的输出数据一般作为 Reducer 的输入数据。

Map 阶段加入 Combiner 之后的数据处理过程可以简述为如下形式。

Map 阶段其他过程：(K1, V1) → list(K2, V2)。

Combine：(K2, list(V2)) → list(K2, V2)。

Reduce 阶段：(K2, list(V2)) → list(K3, V3)。

Combiner 对 Map 阶段来说不是必须使用的，Combiner 主要适用于 Reduce 的输入 Key/Value 与输出 Key/Value 类型完全一致，并且使用 Combiner 不会影响最终输出结果的场景，例如，Combiner 适用于累加求和、获取最大值等应用场景，而不适用于求平均值。

　　下面通过程序来理解 Combiner。

（1）自定义 Combiner。

【例 4-10】WordCountCombiner.java

```
1   /**
2    * 通过Combiner将Key相同的Value进行累加
3    */
4   public class WordCountCombiner extends Reducer<Text, IntWritable, Text, IntWritable> {
5       @Override
6       protected void reduce(Text key, Iterable<IntWritable> values, Context context) throws IOException, InterruptedException {
7           //1.定义一个计数器
8           Integer counter = 0;
9           //2.迭代数组，将输出的K-V对存入context
10          for (IntWritable value : values) {
11              counter += value.get();
12          }
13          context.write(key, new IntWritable(counter));
14      }
15  }
```

（2）实现 Mapper。

【例 4-11】MyMapper.java

```
1   public class MyMapper extends Mapper<LongWritable, Text, Text, IntWritable> {
2       Text word = new Text();
3       IntWritable one = new IntWritable(1);
4       @Override
5       protected void map(LongWritable key, Text value, Context context)
6               throws IOException, InterruptedException {
7           //1.以行为单位，对数据进行处理
8           String line = value.toString();
9           //2.以空格为分隔符，对单词进行拆分
10          String[] words = line.split(" ");
11          //3.迭代数组，将输出的K-V对存入context
12          for (String s : words) {
13              word.set(s);
14              context.write(word, one);
15          }
16      }
17  }
```

（3）实现 Reducer。

【例 4-12】MyReducer.java

```
1   public class MyReducer extends Reducer<Text, IntWritable, Text, IntWritable> {
2       @Override
3       protected void reduce(Text key, Iterable<IntWritable> values, Context context) throws IOException, InterruptedException {
4           //1.定义一个计数器
```

```
5           Integer counter = 0;
6           //2.迭代数组,将输出的 K-V 对存入 context
7           for (IntWritable value : values) {
8               counter += value.get();
9           }
10          context.write(key, new IntWritable(counter));
11      }
12  }
```

(4)创建 MapReduce 作业。

【例 4-13】 CombinerApp.java

```
1   public class CombinerApp {
2       public static void main(String[] args) throws Exception {
3           if (args == null || args.length < 2) {
4               throw new Exception("参数不足,需要两个参数!");
5           }
6           //1.新建配置对象,为配置对象设置文件系统
7           Configuration conf = new Configuration();
8           conf.set("fs.defaultFS", "hdfs://192.168.142.131:9000");
9           //2.设置 Job 属性
10          Job job = Job.getInstance(conf, "CombinerApp");
11          job.setJarByClass(CombinerApp.class);
12          //3.设置数据输入路径
13          Path inPath = new Path(args[0]);
14          FileInputFormat.addInputPath(job, inPath);
15          //4.设置 Job 执行的 Mapper 类
16          job.setMapperClass(MyMapper.class);
17          //5.设置 Combiner
18          job.setCombinerClass(WordCountCombiner.class);
19          //6.设置 Job 执行的 Reducer 类和输出 K-V 类型
20          job.setReducerClass(MyReducer.class);
21          job.setOutputKeyClass(Text.class);
22          job.setOutputValueClass(IntWritable.class);
23          //7.递归删除输出目录
24          FileSystem.get(conf).delete(new Path(args[1]), true);
25          //8.设置输出的文件数
26          job.setNumReduceTasks(3);
27          //9.设置数据输出路径
28          Path outPath = new Path(args[1]);
29          FileOutputFormat.setOutputPath(job, outPath);
30          //10.MapReduce 作业完成后退出系统
31          System.exit(job.waitForCompletion(true) ? 0 : 1);
32      }
33  }
```

(5)使用程序生成一个 HDFS 文件 combinertestdata.txt,该文件包含 5000 条数据。

【例 4-14】 CombinerTestData.java

```
1   public class CombinerTestData {
2       public static void main(String[] args) throws IOException {
3           //1.新建配置文件对象
4           Configuration conf = new Configuration();
5           //2.给配置文件设置 HDFS 文件的默认入口
```

```
6         conf.set("fs.defaultFS", "hdfs://192.168.142.131:9000");
7         //3.通过传入的配置参数得到 FileSystem
8         FileSystem fs = FileSystem.get(conf);
9         //4.设置 HDFS 上存储数据的文件的绝对路径
10        Path p = new Path("/combinertestdata.txt");
11        //5.FileSystem 通过 create()方法获得输出流（FSDataOutputStream）
12        FSDataOutputStream fos = fs.create(p, true, 1024);
13        //6.通过输出流将内容写入 txtData.txt 文件
14        for (int i = 0; i < 5000; i++) {
15            fos.writeBytes((1900 + new Random().nextInt(10) * 10 + new Random().nextInt(10)) + " " + (new Random().nextInt(42 - (-42) + 1) + (-42)) + "\n");
16        }
17        //7.关闭输出流
18        fos.close();
19    }
20 }
```

（6）在 Hadoop 集群上运行 MapReduce 作业。

```
[root@qf01 ~]# hadoop jar testMapReduce-1.0-SNAPSHOT.jar com.qf.mr.combiner.CombinerApp /combinertestdata.txt /outdata/combiner
```

以下是运行作业时，使用 Combiner 的情形。

【例 4-15】 部分运行数据

```
1   Map-Reduce Framework
2       Map input records=5000
3       Map output records=10000
4       Map output bytes=81398
5       Map output materialized bytes=1906
6       Input split bytes=102
7       Combine input records=10000
8       Combine output records=185
9       Reduce input groups=185
10      Reduce shuffle bytes=1906
11      Reduce input records=185
12      Reduce output records=185
```

以下是运行作业时，未使用 Combiner 的情形。
具体做法是将例 4-13 的第 18 行注释掉。

```
//job.setCombinerClass(WordCountCombiner.class);
```

【例 4-16】 部分运行数据

```
1   Map-Reduce Framework
2       Map input records=5000
3       Map output records=10000
4       Map output bytes=81398
5       Map output materialized bytes=101416
6       Input split bytes=102
7       Combine input records=0
8       Combine output records=0
9       Reduce input groups=185
10      Reduce shuffle bytes=101416
11      Reduce input records=10000
12      Reduce output records=185
```

（7）对比例 4-15 和例 4-16 可知，使用 Combiner 时，Reduce 输入记录的条数为 185，未使用 Combiner 时，Reduce 输入记录的条数为 10000。可见使用 Combiner 能够减少 Map 阶段和 Reduce 阶段之间网络传输的数据量。

4.3 MapReduce 作业解析

4.3.1 MapReduce 作业简介

MapReduce 作业（Job）是指一个 MapReduce 应用程序（Application），也就是用户提交的一个计算请求。

在 Java 程序中，Job 是用户作业与 ResourceManager 交互的主要的类。Job 类提供了提交作业、跟踪进度、访问组件任务报告和日志、获取 MapReduce 集群状态信息的工具。作业提交流程如下。

（1）检查作业的输入和输出格式。
（2）计算作业的 InputSplit 值。
（3）将作业的 jar 包和配置文件复制到 FileSystem 上的 MapReduce 系统目录。
（4）将作业提交到 ResourceManager。

作业输出的目的地通常是分布式文件系统（如 HDFS），而输出又可以用作下一个作业的输入，这意味着作业成功或失败完全取决于客户端。在这种情况下，作业的控制主要通过以下两个方法来实现。

（1）Job.submit()：将作业提交到集群并立即返回。
（2）Job.waitForCompletion(boolean)：将作业提交到集群并等待作业完成。

4.3.2 MapReduce 作业运行时的资源调度

MapReduce 作业运行时的资源调度流程，如图 4.6 所示。

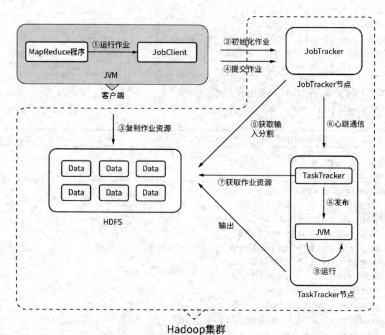

图 4.6　MapReduce 作业运行时的资源调度流程

由图 4.6 可知，MapReduce 作业运行时的资源调度流程主要涉及以下 5 个模块。

（1）客户端（Client）：提交 MapReduce 作业。

（2）YARN 的 ResourceManager：调度各个 NodeManager 上的资源。

（3）YARN 的 NodeManager：启动 Container 运行任务，上报节点资源使用情况、Container 运行情况给 ApplicationMaster。

（4）MapReduce 的 ApplicationMaster：运行 MapReduce 作业。

（5）分布式文件系统（HDFS 等）：存储作业文件。

4.3.3 MapReduce 作业运行流程

MapReduce 作业运行流程（数据处理流程）主要包含两个阶段：Map 阶段、Reduce 阶段。

（1）Map 阶段：FileInputFormat→InputSplit→RecordReader→Mapper→Partition→Sort→Combiner。

（2）Reduce 阶段：Reducer→FileOutputFormat→RecordWriter。

4.4 MapReduce 工作原理

4.4.1 Map 任务工作原理

程序根据 InputFormat 将输入文件分割成 Split，Split 作为 Map 任务的输入数据，每个 Map 任务都有一个内存缓冲区，输入数据经过 Map 阶段处理后，中间结果会写入内存缓冲区，当写入的数据量到达内存缓冲区的阈值时，会启动一个线程将内存中的数据溢写入磁盘，同时不影响 Map 中间结果继续写入内存缓冲区。在溢写过程中，MapReduce 框架会对键进行排序，如果中间结果比较大，会形成多个溢写文件，最后合并所有的溢写文件为一个文件。

4.4.2 Reduce 任务工作原理

每个 Map 任务完成后会形成一个最终文件，该文件按区划分。Reduce 任务启动之前，Map 任务会启动线程来拉取 Map 结果数据到相应的 Reduce 任务，并合并数据，为 Reduce 的数据输入做准备。所有的 Map 任务完成后，数据也拉取合并完毕，Reduce 任务启动，最终将输出结果存入 HDFS。

4.5 Shuffle 阶段

4.5.1 Shuffle 的概念

Shuffle 的本义是洗牌、混洗，是把一组有一定规则的数据尽量转换成一组无规则的数据，越随机越好。MapReduce 中的 Shuffle 更像是洗牌的逆过程，把一组无规则的数据尽量转换成一组具有一定规则的数据。Shuffle 过程，如图 4.7 所示。

由图 4.7 可知，MapReduce 的 Shuffle 阶段主要是指 Mapper 之后到 Reducer 之前的过程，MapReduce 的 Shuffle 可以分为 Map 端的 Shuffle 和 Reduce 端的 Shuffle。

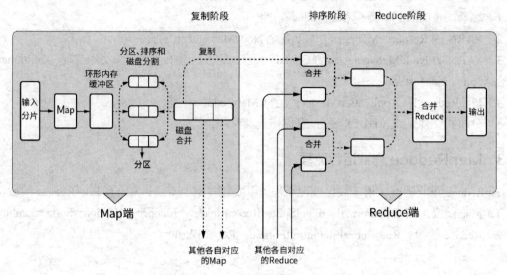

图 4.7　MapReduce 的 Shuffle 阶段

4.5.2　Map 端的 Shuffle

1. 分区

Map 端处理完数据后，键/值被写入缓冲区之前，都会被序列化为字节流。MapReduce 提供 Partitioner 接口，它的作用就是根据键或值及 Reduce 的数量来决定当前的这对输出数据最终应该交由哪个 Reduce 任务处理（分区）。默认对键取 hashCode（哈希码）值，然后再以 Reduce 任务数量取模。默认的取模方式只是为了平衡 Reduce 的处理能力，如果用户自己对 Partitioner 有需求，可以定制并设置到 Job 上。

2. 环形缓冲区

Map 在内存中有一个环形缓冲区（字节数组实现），用于存储任务的输出，默认容量是 100MB，其中 80% 用于缓存。当这部分容量用完时，会启动一个溢出线程进行溢出操作，写入磁盘形成溢写（Spill）文件；在溢出的过程中缓冲区剩余的 20% 对新产生的数据继续缓存（简单来说就是边读边写）。但如果在此期间缓冲区被填满，Map 会阻塞直到写磁盘过程完成。

3. 溢写与排序

缓冲区的数据写入磁盘前，会进行二次快速排序，首先根据数据所属的 partition（分区）排序，然后每个 partition 中再根据键排序。输出数据包括一个索引文件和数据文件。如果设定了 Combiner，Combiner 将在排序输出的基础上运行。Combiner 就是一个简单的 Reduce 操作，它在执行 Map 任务的节点上运行，先对 Map 的输出做一次简单的 Reduce，使得 Map 的输出更紧凑，写入磁盘和传送到 Reducer 的数据更简洁。临时文件会在 Map 任务结束后删除。

4. 文件合并

每次溢写会在磁盘上生成一个溢写文件，如果 Map 的输出量很大，有多次这样的溢写发生，磁盘上就会有多个溢写文件存在。因为最终的文件只有一个，所以需要将这些溢写文件归并到一起，这个过程就叫作文件合并（Merge）。此时可能也有相同的键存在，在这个过程中如果 Client 设置过 Combiner，Combiner 会合并相同的键。

4.5.3 Reduce 端的 Shuffle

1. 复制

Reduce 端默认有 5 个数据复制线程从 Map 端复制数据，其通过 HTTP 方式得到 Map 对应分区的输出文件。Reduce 端并不是等 Map 端执行完将结果传来，而是直接去 Map 端复制（Copy）输出文件（主动拉取数据）。

2. 合并

Reduce 端的 Shuffle 也有一个环形缓冲区，它的容量比 Map 端的灵活。Copy 阶段获得的数据会存放到这个缓冲区中，同样，数据量到达阈值时会发生溢写到磁盘操作。这个过程中如果设置了 Combiner，Combiner 也会执行。这个过程会一直持续到所有的 Map 输出都被复制过来，如果形成了多个磁盘文件，也会进行合并，最后一次合并的结果作为 Reduce 的输入，而不是写入磁盘。

3. 执行

Reducer 的输入文件确定后，整个 Shuffle 操作才最终结束。这时就开始 Reducer 的执行了。最后 Reducer 会把结果存到 HDFS 上。

4.6 优化——数据倾斜

数据倾斜是指处理数据时，大量数据被分配到一个分区中，导致单个节点忙碌，其他节点空闲的问题。数据倾斜明显背离了 MapReduce 并行计算的初衷，降低了处理数据的效率。

数据倾斜是 MapReduce 中常见的问题，解决数据倾斜问题的方法主要有以下几种。

1. 自定义分区

自定义分区中比较常用的是自定义随机分区，核心代码如下。

```
public class RandomPartition extends Partitioner<Text, IntWritable> {
    @Override
    public int getPartition(Text text, IntWritable intWritable, Int numPartitions) {
        Random r = new Random();
        return r.nextInt(numPartitions);
    }
}
```

2. 重新设计 Key

重新设计 Key 是指在 Mapper 中给 Key 加上一个随机数，附带随机数的 Key 不会被大量分配到同一个节点，数据传送到 Reducer 后再把随机数去掉即可。

3. 使用 Combiner

如果 Map 阶段使用了 Combiner，可以选择性地把大量 Key 相同的数据先进行合并，然后再传递给 Reduce 阶段来处理，这样就降低了 Map 阶段向 Reduce 阶段发送的数据量，有利于预防数据倾斜。

4. 增加 Reduce 任务数

只存在一个 Value 的数据比较多，某些数据的 Value 的记录数远远多于其他数据的 Value 的记录数，但是这些数据在总数据量中的占比小于百分之一，这种情况下，容易出现大量 Key 相同的数据被分配到同一个分区的现象，如果 Reduce 任务数较少，某个 Reduce 任务可能会在一个节点上处理大量的数据。这时，可以通过增加 Reduce 任务数，来增加参与处理数据的节点，从而减少数据倾斜。

5. 增加虚拟机内存

只存在一个 Value 的数据非常少（少于几千条），并且极少数据的 Value 有非常多的记录数，这种情况下，可以通过增加虚拟机内存来提高运行效率，从而减少数据倾斜。

4.7 MapReduce 典型案例——排序

Sort（排序）是 MapReduce 中的重要技术，也是程序员的工作重点之一。先来了解 Sort 在执行 Map 任务的过程中所处的位置。

每个 Map 任务在内存中都有一个缓冲区，Map 任务的输出数据会以序列化的形式存储到缓冲区中。该缓冲区的默认大小是 100MB，可以通过 mapreduce.task.io.sort.mb 属性更改缓冲区大小。当缓冲区中的数据量达到特定的阈值时，系统会对缓冲区的数据进行预排序，并溢出数据到磁盘的一个临时文件中。特定的阈值是 mapreduce.task.io.sort.mb 与 mapreduce.map.sort.spill.percent 的乘积，mapreduce.map.sort.spill.percent 默认为 0.80。如果在溢出过程中任一缓冲区完全填满，则执行 Map 任务的线程阻塞。所有的 Map 任务完成后，剩余的数据都会写入磁盘，并且与所有的临时文件合并为一个文件。

MapReduce 中常见的排序方式有以下 3 种。

4.7.1 部分排序

部分排序是指 MapReduce 中对每个 Reduce 任务的输出 Key 进行排序，这是 MapReduce 默认的排序方式。下面通过编写程序来理解部分排序，具体步骤如下。

（1）在 IntelliJ IDEA 中 testMapReduce 模块下新建包 com.qf.mr.sort.partialsort。

（2）实现 Mapper 和 Reducer。使用 com.qf.mr.wordcount 包下的 MyMapper 和 MyReducer 即可，不需要再写程序。

（3）创建 MapReduce 作业。

【例 4-17】PartialSortApp.java

```
1   public class PartialSortApp {
2       public static void main(String[] args) throws Exception {
3           if (args == null || args.length < 2) {
4               throw new Exception("参数不足,需要两个参数!");
5           }
6           //1.新建配置对象，为配置对象设置文件系统
7           Configuration conf = new Configuration();
8           conf.set("fs.defaultFS", "hdfs://192.168.142.131:9000");
9           //2.设置Job属性
10          Job job = Job.getInstance(conf, "WordCountApp");
11          job.setJarByClass(PartialSortApp.class);
12          //3.设置数据输入路径
13          Path inPath = new Path(args[0]);
14          FileInputFormat.addInputPath(job, inPath);
15          //4.设置Map阶段的属性
16          job.setMapperClass(MyMapper.class);
17          job.setMapOutputKeyClass(Text.class);
```

```
18          job.setMapOutputValueClass(IntWritable.class);
19          //5.设置 Reduce 阶段的属性
20          job.setReducerClass(MyReducer.class);
21          job.setOutputKeyClass(Text.class);
22          job.setOutputValueClass(IntWritable.class);
23          //设置 reduce 个数
24          job.setNumReduceTasks(5);
25          //6.设置数据输出路径
26          Path outPath = new Path(args[1]);
27          FileOutputFormat.setOutputPath(job, outPath);
28          //7.MapReduce 作业完成后退出系统
29          System.exit(job.waitForCompletion(true) ? 0 : 1);
30      }
31  }
```

（4）在 Hadoop 集群上运行 MapReduce 作业。

用 Maven Projects 将模块 testMapReduce 打成 Jar 包，将 testMapReduce-1.0-SNAPSHOT.jar 发送到活跃虚拟机的/root 目录下，运行 MapReduce 作业。

```
[root@qf01 ~]# hadoop jar testMapReduce-1.0-SNAPSHOT.jar com.qf.mr.sort.partialsort.PartialSortApp /word.txt /outdata/partialsort
```

（5）在活跃的虚拟机上，查看 MapReduce 作业的运行结果。

```
[root@qf01 ~]# hdfs dfs -cat /outdata/partialsort/part-r-00000
hi          3
mapreduce   1
qianfeng 2
[root@qf01 ~]# hdfs dfs -cat /outdata/partialsort/part-r-00001
[root@qf01 ~]# hdfs dfs -cat /outdata/partialsort/part-r-00002
[root@qf01 ~]# hdfs dfs -cat /outdata/partialsort/part-r-00003
hadoop  1
hello   2
world   1
[root@qf01 ~]# hdfs dfs -cat /outdata/partialsort/part-r-00004
```

由以上运行结果可知，只有 part-r-00000 和 part-r-00003 两个文件存放了数据，两个文件中的数据都按照 Key 的字典序进行了排序，但是将两个文件合并到一起后，并不能保证数据是全局有序的。在只需要对 Key 排序的应用场景中，对数据进行部分排序已经足够了。

4.7.2 全排序

全排序是指对所有 Reduce 任务输出的所有 Key 进行排序。对数据进行全排序，主要有 3 种方法：使用一个 Partitioner，自定义 Partitioner，对 Key 进行采样。

（1）使用一个 Partitioner，即只执行一个 Reduce 任务。该方法的缺点比较明显，当处理大型数据集时，一台服务器需要处理所有的数据，MapReduce 的分布式并行计算的优势无法体现出来。

（2）自定义 Partitioner。由自定义 Partitioner 的 MapReduce 作业运行结果可知，自定义 Partitioner 实现了对数据的全排序。这种方法需要遍历整个数据集，当处理大数据集时，直接自定义 Partitioner 难度很大，因此在实际开发环境中，通常使用对 Key 进行采样的方法。

（3）对 Key 进行采样。采样是在样本中查看一小部分 Key，得到 Key 的近似分布，目的是据此自定义 Partitioner。MapReduce 提供了 3 个采样器。

- 随机采样器（RandomSampler）：以指定的采样率均匀地从一个数据集中取得样本。
- 间隔采样器（IntervalSampler）：适用于 Key 有序的场景，以相等的间隔对样本进行采样。
- 切片采样器（SplitSampler）：只对每个切片的前 *n* 条数据进行采样。

从采样效率来看，切片采样器效率最高，间隔采样器效率次之，随机采样器效率最低。

在实际生产环境中，通常对序列文件使用采样器来实现全排序，下面结合程序来理解 3 个采样器。

创建 SequenceFile 源文件。

【例 4-18】 CreateSequenceFile.java

```
1   public class CreateSequenceFile {
2       public static void main(String[] args) throws Exception {
3           //1.指定文件系统
4           Configuration conf = new Configuration();
5           conf.set("fs.defaultFS", "hdfs://192.168.142.131:8020");
6           //2.设置文件输出路径
7           Path outputPath = new Path("/SamplerSequenceFile.seq");
8           //3.指定写操作的路径和内容
9           SequenceFile.Writer.Option optPath = SequenceFile.Writer.file(outputPath);
10          SequenceFile.Writer.Option optKey = SequenceFile.Writer.keyClass(IntWritable.class);
11          SequenceFile.Writer.Option optVal = SequenceFile.Writer.valueClass(Text.class);
12          //4.执行写操作
13          SequenceFile.Writer writer = SequenceFile.createWriter(conf, optPath, optKey, optVal);
14          //5.向文件 MySequenceFile.seq 添加内容
15          IntWritable key = new IntWritable();
16          Text value = new Text();
17          for (int i = 0; i < 1000; i++) {
18              key.set(i);
19              value.set("Tom" + i);
20              writer.append(key, value);
21          }
22          writer.close();
23      }
24  }
```

实现 Mapper。

【例 4-19】 MyMapper.java

```
1   public class MyMapper extends Mapper<IntWritable, Text, IntWritable, Text> {
2       @Override
3       protected void map(IntWritable key, Text value, Context context) throws IOException, InterruptedException {
4           context.write(key, value);
5       }
6   }
```

创建 MapReduce 作业。

【例 4-20】 TestSamplerApp.java

```
1   public class TestSamplerApp {
2       public static void main(String[] args) throws Exception {
```

```java
3          if (args == null || args.length < 2) {
4              throw new Exception("参数不足,需要两个参数!");
5          }
6          //1.新建配置对象,为配置对象设置文件系统
7          Configuration conf = new Configuration();
8          conf.set("fs.defaultFS", "hdfs://192.168.142.131:9000");
9          //2.设置 Job 属性
10         Job job = Job.getInstance(conf, "test sampler");
11         job.setJarByClass(TestSamplerApp.class);
12         //3.设置数据输入路径
13         Path inPath = new Path(args[0]);
14         FileInputFormat.addInputPath(job, inPath);
15         //4.设置 Job 的输入格式
16         job.setInputFormatClass(SequenceFileInputFormat.class);
17         //5.设置 Job 执行的 Mapper 类
18         job.setMapperClass(MyMapper.class);
19         //6.使用 MapReduce 内置的全排序分区类
20         job.setPartitionerClass(TotalOrderPartitioner.class);
21         //7.指定全排序分区文件的位置
22         TotalOrderPartitioner.setPartitionFile(job.getConfiguration(), new Path
("/PartitionFile.seq"));
23         //8.设置 Job 的输出 Key 和 Value
24         job.setOutputKeyClass(IntWritable.class);
25         job.setOutputValueClass(Text.class);
26         //9.设置输出的文件数
27         job.setNumReduceTasks(3);
28         //递归删除输出目录,避免出现因多次运行而产生"输出目录已存在"的异常
29         FileSystem.get(conf).delete(new Path(args[1]), true);
30         //10.设置数据输出路径
31         Path outPath = new Path(args[1]);
32         FileOutputFormat.setOutputPath(job, outPath);
33         //11.设置采样器
34         /**
35          * (1)随机采样器
36          * RandomSampler(0.1, 500, 3)中的三个参数含义如下
37          *     freq: 选中每个 Key 的概率
38          *     numSamples: 从所有选定的切片中获得的样本总数
39          *     maxSplitsSampled: 文件最大切片数
40          */
41         InputSampler.Sampler sampler = new InputSampler.RandomSampler(0.1, 500, 3);
42         /**
43          * (2)间隔采样器
44          * IntervalSampler(0.1, 3)中的两个参数分别是 freq、maxSplitsSampled
45          */
46         //InputSampler.Sampler sampler = new    InputSampler.IntervalSampler(0.1,
3);
47         /**
48          * (3)切片采样器
49          * SplitSampler(500, 3)中的两个参数分别是 numSamples、maxSplitsSampled
50          */
```

```
51          //InputSampler.Sampler sampler = new InputSampler.SplitSampler(500, 3);
52          //12.写入分区文件
53          InputSampler.writePartitionFile(job, sampler);
54          //13.MapReduce 作业完成后退出系统
55          System.exit(job.waitForCompletion(true) ? 0 : 1);
56      }
57  }
```

在 Hadoop 集群上运行 MapReduce 作业。

```
[root@qf01 ~]# hadoop jar testMapReduce-1.0-SNAPSHOT.jar com.qf.mr.sort.
totalsort.sampler.TestSamplerApp /SamplerSequenceFile.seq /outdata/totalsortsampler
```

查看 MapReduce 作业的运行结果如下。

```
[root@qf01 ~]# hdfs dfs -cat /outdata/totalsortsampler/part-r-00000
0       Tom0
1       Tom1
2       Tom2
...
290     Tom290
291     Tom291
292     Tom292
[root@qf01 ~]# hdfs dfs -cat /outdata/totalsortsampler/part-r-00001
293     Tom293
294     Tom294
295     Tom295
...
639     Tom639
640     Tom640
641     Tom641
[root@qf01 ~]# hdfs dfs -cat /outdata/totalsortsampler/part-r-00002
642     Tom642
643     Tom643
644     Tom644
...
997     Tom997
998     Tom998
999     Tom999
```

由 MapReduce 作业的运行结果可知，采样器生成两个 Key，把所有的 Key 分成 3 段（0~292，293~641，642~999），并且该 3 段数据按照 Key 进行了全排序，被存放到了 3 个输出文件中。由于使用了随机采样器，以上结果是随机的，所以读者得到的结果很可能与以上结果不同。

采样器生成的两个 Key 被存放到了 HDFS 的/PartitionFile.seq 文件中，下面通过程序查看 PartitionFile.seq 的内容。

【例 4-21】SequenceFileReader.java

```
1   public class SequenceFileReader {
2       public static void main(String[] args) throws Exception {
3           //1.指定文件系统
4           Configuration conf = new Configuration();
5           conf.set("fs.defaultFS", "hdfs://192.168.142.131:9000");
6           //2.设置文件读取路径
```

```
7          Path path = new Path("/PartitionFile.seq");
8          //3.指定读操作的路径
9          Option optPath = SequenceFile.Reader.file(path);
10         //4.执行读操作
11         SequenceFile.Reader reader = new SequenceFile.Reader(conf, optPath);
12         //5.将位置信息、写操作时写入的数据读取出来
13         Writable key = (Writable) ReflectionUtils.newInstance(reader.getKeyClass(), conf);
14         Writable value = (Writable) ReflectionUtils.newInstance(reader.getValueClass(), conf);
15         long position = reader.getPosition();
16         while (reader.next(key, value)) {
17             String syncSeen = reader.syncSeen() ? "*" : "";
18             System.out.printf("[%s%s]\t%s\t%s\n", position, syncSeen, key, value);
19             position = reader.getPosition();
20         }
21         reader.close();
22     }
23 }
```

以上程序的运行结果如下。

```
[136]    407   (null)
[156]    702   (null)
```

由程序运行结果可知，确定分区边界的两个 Key 分别是 407，702。

以上实现了通过采样对所有 Key 进行全排序。

在实际开发环境中，排序是一项重要的任务。对于大数据开发工程师而言，写出好的排序算法是一件相对困难的事情。但幸运的是，MapReduce 框架已经提供了大多数常用的排序算法，开发工程师只需要提供基本的排序规则即可。

4.8 MapReduce 典型案例——倒排索引

倒排索引（Inverted Index），又称反向索引，它被用于存储某个单词在一个文档或者一组文档中存储位置的映射。倒排索引主要通过单词来查找包含单词的文档，而不是通过文档来查找单词，因此被称为倒排索引。

倒排索引是文档检索系统中最常用的索引方法，现代搜索引擎都是基于倒排索引。下面通过案例来理解倒排索引。

案例：做一个倒排索引，统计某个单词在各个文件中的出现次数。

4.8.1 准备模拟数据

在活跃的虚拟机上新建需要实现倒排索引的 3 个文件 index.txt、hadoop-info.txt、spark-info.txt，并填入相应的模拟数据（单词间的分隔符为空格）。

index.txt 文件数据如下。

```
Hadoop is good Hadoop is nice
```

hadoop-info.txt 文件数据如下。

```
Hadoop is better
```

spark-info.txt 文件数据如下。

```
Spark is good Spark is nice
```

4.8.2 输出数据解析

目标输出数据样式如下。

```
Hadoop  hadoop-info.txt,1;index.txt,2
```

其中，Hadoop 是在文件中出现的单词，hadoop-info.txt 和 index.txt 是含有 Hadoop 一词的文件，1 和 2 分别是 Hadoop 一词出现在 hadoop-info.txt 和 index.txt 文件中的次数。

4.8.3 编写 MapReduce 程序

自定义 Combiner。

【例 4-22】InvertedIndexCombiner.java

```
1    public class InvertedIndexCombiner extends Reducer<Text, Text, Text, Text> {
2        @Override
3        protected void reduce(Text key, Iterable<Text> values, Context context) throws
IOException, InterruptedException {
4            int count = 0;
5            for (Text value : values) {
6                count += Integer.parseInt(value.toString());
7            }
8            String[] keySplit = key.toString().split("_");
9            context.write(new Text(keySplit[0]), new Text(keySplit[1] + "," + count +
";"));
10       }
11   }
```

实现 Mapper。

【例 4-23】MyMapper.java

```
1    public class MyMapper extends Mapper<Object, Text, Text, Text> {
2        @Override
3        protected void map(Object key, Text value, Context context)
4                throws IOException, InterruptedException {
5            //1.先获取文件名字
6            InputSplit is = context.getInputSplit();
7            String fileName = ((FileSplit) is).getPath().getName();
8            //2.以行为单位，对数据进行处理
9            String line = value.toString();
10           //3.以空格为分隔符，对单词进行拆分
11           String[] words = line.split(" ");
12           //4.迭代数组，将输出的 K-V 对存入 context
13           for (String word : words) {
14               context.write(new Text(word + "_" + fileName), new Text( "1"));
15           }
16       }
17   }
```

实现 Reducer。

【例 4-24】 MyReducer.java

```
1    public class MyReducer extends Reducer<Text, Text, Text, Text> {
2        @Override
3        protected void reduce(Text key, Iterable<Text> values, Context context) throws IOException, InterruptedException {
4            String result = "";
5            for (Text value : values) {
6                result += value.toString();
7            }
8            context.write(key, new Text(result.substring(0, result.length() - 1)));
9        }
10   }
```

创建 MapReduce 作业。

【例 4-25】 InvertedIndexApp.java

```
1    public class InvertedIndexApp {
2        public static void main(String[] args) throws Exception {
3            if (args == null || args.length < 2) {
4                throw new Exception("参数不足,需要两个参数!");
5            }
6            //1.新建配置对象,为配置对象设置文件系统
7            Configuration conf = new Configuration();
8            conf.set("fs.defaultFS", "hdfs://192.168.142.131:9000");
9            //2.设置Job属性
10           Job job = Job.getInstance(conf, "InvertedIndexApp");
11           job.setJarByClass(InvertedIndexApp.class);
12           //3.设置数据输入路径
13           Path inPath = new Path(args[0]);
14           FileInputFormat.addInputPath(job, inPath);
15           //4.设置Job执行的Mapper类
16           job.setMapperClass(MyMapper.class);
17           //5.设置Combiner
18           job.setCombinerClass(InvertedIndexCombiner.class);
19           //6.设置Job执行的Reducer类和输出K-V类型
20           job.setReducerClass(MyReducer.class);
21           job.setOutputKeyClass(Text.class);
22           job.setOutputValueClass(Text.class);
23           //7.递归删除输出目录
24           FileSystem.get(conf).delete(new Path(args[1]), true);
25           //8.设置数据输出路径
26           Path outPath = new Path(args[1]);
27           FileOutputFormat.setOutputPath(job, outPath);
28           //9.MapReduce作业完成后退出系统
29           System.exit(job.waitForCompletion(true) ? 0 : 1);
30       }
31   }
```

在活跃的虚拟机的 HDFS 上新建目录/invertedindex_in,将 index.txt、hadoop-info.txt、spark-info.txt 上传到该目录下。

```
[root@qf01 ~]# hdfs dfs -mkdir /invertedindex_in
[root@qf01 ~]# hdfs dfs -put index.txt hadoop-info.txt spark-info.txt /invertedindex_in
```

在 Hadoop 集群上运行 MapReduce 作业。

```
[root@qf01 ~]# hadoop jar testMapReduce-1.0-SNAPSHOT.jar com.qf.mr.invertedindex.
InvertedIndexApp /invertedindex_in /outdata/invertedindex
```

MapReduce 作业运行结果如下。

```
Hadoop   hadoop-info.txt,1;index.txt,2
Spark    spark-info.txt,2
better   hadoop-info.txt,1
good     spark-info.txt,1;index.txt,1
is       index.txt,2;hadoop-info.txt,1;spark-info.txt,2
nice     spark-info.txt,1;index.txt,1
```

4.9 MapReduce 典型案例——连接

在实际开发工作中，经常会从多个数据源获取海量数据，将这些数据放到一起进行处理时，经常会用到 Join（连接）操作。下面通过案例来理解 Join。

案例：使用 Join 操作，将两个文件中的人名和城市对应，放在一个文件中。

4.9.1 准备模拟数据

在活跃的虚拟机上新建需要进行 Join 操作的 2 个文件 name.txt、city.txt，并填入相应的模拟数据（单词间的分隔符为空格）。

name.txt 文件数据样式为"姓名 城市编号"，具体数据如下。

```
Sophie 1
Cocona 1
Tom 2
Jack 3
Lucy 3
Rose 2
```

city.txt 文件数据样式为"城市编号 城市名称"，具体数据如下。

```
1 beijing
2 shanghai
3 shenzhen
```

4.9.2 输出数据解析

目标输出数据样式为"姓名 城市名称"。

```
Cocona   beijing
```

4.9.3 编写 MapReduce 程序

实现 Mapper。

【例 4-26】 MyMapper.java

```java
public class MyMapper extends Mapper<Object, Text, Text, Text> {
    @Override
    protected void map(Object key, Text value, Context context)
            throws IOException, InterruptedException {
        //1.以行为单位，对数据进行处理
        String line = value.toString();
        //2.切分一行文本
        StringTokenizer words = new StringTokenizer(line);
        int i = 0;
        String tempKey = new String();
        String tempValue = new String();
        String filetype = new String();
        //3.迭代处理切分出的单词
        while (words.hasMoreTokens()) {
            String word = words.nextToken();
            //为两个文件 name.txt 和 city.txt 设置编号
            if (word.charAt(0) >= '0' && word.charAt(0) <= '9') {
                tempKey = word;
                if (i > 0) {
                    filetype = "1";
                } else {
                    filetype = "2";
                }
                continue;
            }
            tempValue += word + " ";
            i++;
        }
        // 4.将输出的 K-V 对存入 context
        context.write(new Text(tempKey), new Text(filetype + "+" + tempValue));
    }
}
```

实现 Reducer。

【例 4-27】 MyReducer.java

```java
public class MyReducer extends Reducer<Text, Text, Text, Text> {
    @Override
    protected void reduce(Text key, Iterable<Text> values, Context context) throws IOException, InterruptedException {
        int cityIDFromName = 0;
        String[] name = new String[10];
        int cityID = 0;
        String[] city = new String[10];
        Iterator value = values.iterator();
        while (value.hasNext()) {
            String record = value.next().toString();
            int len = record.length();
            int i = 2;
            if (0 == len) {
                continue;
            }
            //获取两个文件的编号
```

```
17              char filetype = record.charAt(0);
18              if ('1' == filetype) {
19                  name[cityIDFromName] = record.substring(i);
20                  cityIDFromName++;
21              }
22              if ('2' == filetype) {
23                  city[cityID] = record.substring(i);
24                  cityID++;
25              }
26          }
27          //求笛卡尔积
28          if (0 != cityIDFromName && 0 != cityID) {
29              for (int j = 0; j < cityIDFromName; j++) {
30                  for (int k = 0; k < cityID; k++) {
31                      context.write(new Text(name[j]),
32                              new Text(city[k]));
33                  }
34              }
35          }
36      }
37  }
```

创建 MapReduce 作业。

【例 4-28】JoinApp.java

```
1   public class JoinApp {
2       public static void main(String[] args) throws Exception {
3           if (args == null || args.length < 2) {
4               throw new Exception("参数不足,需要两个参数!");
5           }
6           //1.新建配置对象，为配置对象设置文件系统
7           Configuration conf = new Configuration();
8           conf.set("fs.defaultFS", "hdfs://192.168.142.131:9000");
9           //2.设置 Job 属性
10          Job job = Job.getInstance(conf, "JoinApp");
11          job.setJarByClass(JoinApp.class);
12          //3.设置数据输入路径
13          Path inPath = new Path(args[0]);
14          FileInputFormat.addInputPath(job, inPath);
15          //4.设置 Job 执行的 Mapper 类
16          job.setMapperClass(MyMapper.class);
17          //5.设置 Job 执行的 Reducer 类和输出 K-V 类型
18          job.setReducerClass(MyReducer.class);
19          job.setOutputKeyClass(Text.class);
20          job.setOutputValueClass(Text.class);
21          //6.递归删除输出目录
22          FileSystem.get(conf).delete(new Path(args[1]), true);
23          //7.设置数据输出路径
24          Path outPath = new Path(args[1]);
25          FileOutputFormat.setOutputPath(job, outPath);
26          //8.MapReduce 作业完成后退出系统
27          System.exit(job.waitForCompletion(true) ? 0 : 1);
28      }
```

```
 29    }
```

在活跃的虚拟机的 HDFS 上新建目录/join_in,将 name.txt、city.txt 上传到该目录下。

```
[root@qf01 ~]# hdfs dfs -mkdir /join_in
[root@qf01 ~]# hdfs dfs -put name.txt city.txt /join_in
```

在 Hadoop 集群上运行 MapReduce 作业。

```
[root@qf01 ~]# hadoop jar testMapReduce-1.0-SNAPSHOT.jar com.qf.mr.join.JoinApp /join_in /outdata/join
```

MapReduce 作业运行结果如下。

```
Cocona  beijing
Sophie  beijing
Rose    shanghai
Tom     shanghai
Lucy    shenzhen
Jack    shenzhen
```

4.10 MapReduce 典型案例——平均分以及百分比

在实际开发工作中,会遇到求平均分以及百分比的需求。

案例:有一批学生考试成绩数据,求出每人的总平均分,以及每个分数段人数占总人数的百分比。

4.10.1 准备模拟数据

在活跃的虚拟机上新建学生成绩文件 student.txt,并填入相应的模拟数据(单词间的分隔符为空格)。

student.txt 文件数据样式为"姓名 语文成绩 数学成绩 英语成绩",具体数据如下。

```
lh   92 68 70
zyt  94 88 75
ls   96 78 78
hgw  90 70 56
yxx  80 88 73
hz   90 98 70
xyd  60 88 73
hj   90 58 70
cs   50 58 11
```

4.10.2 输出数据解析

目标输出数据样式为"分数段 人数 占总人数的百分比"。

```
<60  1  8%
```

4.10.3 编写 MapReduce 程序

(1)在 Map 中求出每个的人平均分,输出如<"<60",1>。
(2)在 Reduce 中,将每个"分数段"及"人数"放入一个 HashMap(哈希映射),如<"<60",3>。

97

（3）将 HashMap 中的 Value 进行累加，得出总人数。
（4）根据 Value 和总人数计算出"百分比"。
（5）在 cleanup()中将最终结果一次输出。

实现 Mapper。

【例 4-29】MyMapper.java

```java
1   public class MyMapper extends Mapper<LongWritable, Text, Text, Text>{
2       Text k = new Text();
3       Text v = new Text();
4       @Override
5       protected void map(LongWritable key, Text value,Context context)
6               throws IOException, InterruptedException {
7           //一个学生的成绩
8           String line = value.toString();
9           String scores [] = line.split("\t");
10          String chinese = scores[1];
11          String math = scores[2];
12          String english = scores[3];
13          double avg = (Double.parseDouble(chinese) + Double.parseDouble(math)
14              + Double.parseDouble(english))/(scores.length-1);
15          //判断
16          if(avg < 60){
17              k.set("<60");
18              v.set("1");
19          } else if(avg >= 60 && avg < 70){
20              k.set("60-70");
21              v.set("1");
22          } else if(avg >= 70 && avg < 80){
23              k.set("70-80");
24              v.set("1");
25          } else if(avg >= 80 && avg < 90){
26              k.set("80-90");
27              v.set("1");
28          } else if(avg >= 90 && avg <= 100){
29              k.set("90-100");
30              v.set("1");
31          }
32          //context.getConfiguration().setInt("counter", counter);
33          context.write(k, v);
34      }
35  }
```

实现 Reducer。

【例 4-30】MyReducer.java

```java
1   public class MyReducer extends Reducer<Text, Text, Text, Text>{
2       //在 reduce()方法执行之前执行一次(仅一次)
3       @Override
4       protected void setup(Context context)
5               throws IOException, InterruptedException {
6           context.write(new Text("分数段"), new Text("人数"+"\t"+"百分比"));
7       }
8       int totalPerson = 0;
```

```
9              List<String> li = new ArrayList<String>();
10             @Override
11             protected void reduce(Text key, Iterable<Text> value,Context context)
12                 throws IOException, InterruptedException {
13                 /**
14                  * <60 list(1,1)
15                  */
16                 int i = 0;
17                 for (Text t : value) {
18                     if(key.toString().equals("<60")){
19                         //l6 ++;
20                         i ++ ;
21                     } else if (key.toString().equals("60-70")){
22                         //g6l7 ++;
23                         i ++ ;
24                     } else if (key.toString().equals("70-80")){
25                         //g7l8 ++ ;
26                         i ++ ;
27                     } else if (key.toString().equals("80-90")){
28                         //g8l9 ++;
29                         i ++ ;
30                     } else if (key.toString().equals("90-100")){
31                         //g9l10 ++;
32                         i ++ ;
33                     }
34                     totalPerson ++ ;
35                 }
36                 li.add(key.toString()+"_"+i);//输出效果   <"<60  3">
37                 //context.getConfiguration().get("counter");
38             }
39             //在reduce()方法执行之后执行一次(仅一次)
40             @Override
41             protected void cleanup(Context context)
42                 throws IOException, InterruptedException {
43                 for (String s : li) {
44                     String l [] = s.split("_");
45                     context.write(new Text(l[0]), new Text(l[1]+"\t"+Double.parseDouble(l[1])/totalPerson*100+"%"));
46                 }
47             }
48         }
```

创建 MapReduce 作业。

【例 4-31】 AvgApp.java

```
1   public class AvgApp {
2   public static void main(String[] args) throws Exception {
3       if (args == null || args.length < 2) {
4           throw new Exception("参数不足,需要两个参数!");
5       }
6       //1.新建配置对象,为配置对象设置文件系统
7       Configuration conf = new Configuration();
8       conf.set("fs.defaultFS", "hdfs://192.168.142.131:8020");
9       //2.设置Job属性
10      Job job = Job.getInstance(conf, "AvgApp");
```

```
11              job.setJarByClass(AvgApp.class);
12          //3.设置数据输入路径
13              Path inPath = new Path(args[0]);
14              FileInputFormat.addInputPath(job, inPath);
15          //4.设置Job执行的Mapper类
16              job.setMapperClass(MyMapper.class);
17          //5.设置Job执行的Reducer类和输出K-V类型
18              job.setReducerClass(MyReducer.class);
19              job.setOutputKeyClass(Text.class);
20              job.setOutputValueClass(Text.class);
21          //6.递归删除输出目录
22              FileSystem.get(conf).delete(new Path(args[1]), true);
23          //7.设置数据输出路径
24              Path outPath = new Path(args[1]);
25              FileOutputFormat.setOutputPath(job, outPath);
26          //8.MapReduce作业完成后退出系统
27              System.exit(job.waitForCompletion(true) ? 0 : 1);
28          }
29      }
```

在活跃的虚拟机的 HDFS 上新建目录/avg_in,将 student.txt 上传到该目录下。

```
[root@qf01 ~]# hdfs dfs -mkdir /avg_in
[root@qf01 ~]# hdfs dfs -put student.txt /avg_in
```

在 Hadoop 集群上运行 MapReduce 作业。

```
[root@qf01 ~]# hadoop jar testMapReduce-1.0-SNAPSHOT.jar com.qf.mr.avg.AvgApp /avg_in /outdata/avg
```

MapReduce 作业运行结果如下。

```
<60      1   8%
60-70    2   16%
70-80    5   33%
80-90    2   6%
90-100   3   28%
```

4.11 MapReduce 典型案例——过滤敏感词汇

在实际工作中,经常需要去除某文件中不需要的字段,或者一些敏感词汇。
案例:过滤掉一个文档包含的某个或某些敏感词汇。

4.11.1 准备模拟数据

在活跃的虚拟机上新建学生成绩文件 article.txt,并填入相应的模拟数据。
article.txt 文件数据如下。

```
We ask that you please do not send us emails privately asking for support.
   We are non-paid volunteers who help out with the project and we do not necessarily have the time or energy to help people on an individual basis. Instead, we have setup mailing lists for each module which often contain hundreds of individuals who will help answer detailed requests for help. The benefit of using mailing lists over private communication is that it is a shared resource where others can also learn from common mistakes and as a community we all grow together.
```

4.11.2 创建敏感词库

需要过滤的敏感词库 sensitive.txt 内容如下。

```
ask from all
```

4.11.3 编写 MapReduce 程序

实现 Mapper。

【例 4-32】MyMapper.java

```
1    public class MyMapper extends Mapper<LongWritable, Text, Text, Text>{
2        /**
3         * 读取小文件进行缓存 （分布式缓存）
4         */
5        static List<String> li = new ArrayList<String>();
6        @Override
7        protected void setup(Context context)throws IOException, InterruptedException {
8            //获取缓存文件路径的数组
9            Path [] paths = DistributedCache.getLocalCacheFiles(context.getConfiguration());
10           //循环读取每一个缓存文件
11           for (Path p : paths) {
12               //获取文件名字
13               String fileName = p.getName();
14               if(fileName.equals("dir")){
15                   BufferedReader sb = null;
16                   sb = new BufferedReader(new FileReader(new File(p.toString())));
17                   //读取 BufferedReader 里面的数据
18                   String tmp = null;
19                   while ( (tmp = sb.readLine()) != null) {
20                       String ss []= tmp.split(" ");
21                       for (String s : ss) {
22                           li.add(s);
23                       }
24                   }
25                   //关闭 sb 对象
26                   sb.close();
27               }
28           }
29       }
30       @Override
31       protected void map(LongWritable key, Text value,Context context)
32               throws IOException, InterruptedException {
33           String line = value.toString();
34           StringTokenizer lines = new StringTokenizer(line);
35           while (lines.hasMoreTokens()) {
36               //判断每一个单词是否是敏感词汇
37               String word = lines.nextToken();
38               if(!li.contains(word)){
39                   context.write(new Text(word), new Text("1"));
40               }
```

```
41              }
42          }
43          @Override
44          protected void cleanup(Context context)throws IOException, InterruptedException {
45          }
46      }
```

实现 Reducer。

【例 4-33】MyReducer.java

```
1   public class MyReducer extends Reducer<Text, Text, Text, Text>{
2       @Override
3       protected void setup(Context context)throws IOException, InterruptedException {
4       }
5       @Override
6       protected void reduce(Text key, Iterable<Text> value,Context context)
7               throws IOException, InterruptedException {
8           int counter = 0;
9           for (Text t : value) {
10              counter += Integer.parseInt(t.toString());
11          }
12          context.write(key, new Text(counter+""));
13      }
14      protected void cleanup(Context context)throws IOException, InterruptedException {
15      }
16  }
```

创建 MapReduce 作业。

【例 4-34】ArticleApp.java

```
1   public class ArticleApp {
2       public static void main(String[] args) throws Exception {
3           if (args == null || args.length < 2) {
4               throw new Exception("参数不足,需要两个参数!");
5           }
6           //1.新建配置对象,为配置对象设置文件系统
7           Configuration conf = new Configuration();
8           conf.set("fs.defaultFS", "hdfs://192.168.142.131:9000");
9           //2.设置 Job 属性
10          Job job = Job.getInstance(conf, " ArticleApp ");
11          job.setJarByClass(ArticleApp.class);
12          //3.设置数据输入路径
13          Path inPath = new Path(args[0]);
14          FileInputFormat.addInputPath(job, inPath);
15          //4.设置 Job 执行的 Mapper 类
16          job.setMapperClass(MyMapper.class);
17          //5.设置 Job 执行的 Reducer 类和输出 K-V 类型
18          job.setReducerClass(MyReducer.class);
19          job.setOutputKeyClass(Text.class);
20          job.setOutputValueClass(Text.class);
21          //6.递归删除输出目录
22          FileSystem.get(conf).delete(new Path(args[1]), true);
```

```
23              //7.设置数据输出路径
24              Path outPath = new Path(args[1]);
25              FileOutputFormat.setOutputPath(job, outPath);
26              //8.MapReduce 作业完成后退出系统
27              System.exit(job.waitForCompletion(true) ? 0 : 1);
28          }
29      }
```

在活跃的虚拟机的 HDFS 上新建目录/ article_in，将 article.txt 上传到该目录下。

```
[root@qf01 ~]# hdfs dfs -mkdir /article_in
[root@qf01 ~]# hdfs dfs -put article.txt /article_in
```

在 Hadoop 集群上运行 MapReduce 作业。

```
[root@qf01 ~]# hadoop jar testMapReduce-1.0-SNAPSHOT.jar com.qf.mr.article.ArticleApp /article_in /outdata/article
```

MapReduce 作业运行结果如下。

```
We that you please do not send us emails privately asking for support.
We are non-paid volunteers who help out with the project and we do not necessarily have
the time or energy to help people on an individual basis. Instead, we have setup mailing lists
for each module which often contain hundreds of individuals who will help answer detailed
requests for help. The benefit of using mailing lists over private communication is that it
is a shared resource where others can also learn common mistakes and as a community we grow
together.
```

4.12 本章小结

本章主要讲解 Hadoop 的分布式计算框架。首先讲解了 MapReduce 的基本概念，让大家对 Hadoop 分布式计算框架有基本的了解；然后对 MapReduce 的编程组件、作业流程以及工作原理进行了讲解，让大家明白 MapReduce 内部运行机制；最后通过大量的典型案例进行反复练习，使初学者更好地理解分布式计算思想，快速掌握常见的编程模型。

4.13 习题

1. 填空题

（1）MapReduce 把数据处理流程分成两个主要阶段：＿＿＿＿阶段和＿＿＿＿阶段。
（2）由 InputFormat 源码可知，InputFormat 有两个方法：getSplits()和＿＿＿＿。
（3）Map 是＿＿＿＿，负责数据的＿＿＿＿，Reduce 是＿＿＿＿，负责数据的＿＿＿＿。
（4）LineRecordReader 以＿＿＿＿作为 Key，＿＿＿＿作为 Value。
（5）Partitioner 的数量与＿＿＿＿任务数相同。

2. 选择题

（1）MapReduce 的特点有（　　）。

    A．易于编程   B．具有良好的扩展性
    C．具有高容错性   D．适合 PB 级以上海量数据的离线处理

（2）作业的控制主要通过（　　）方法来实现。
　　A. submit()　　　　　　　　　　　B. getInstance()
　　C. waitForCompletion(boolean)　　D. setJarByClass()
（3）MyMapper 类中输出的 K-V 对类型需要和 MyReducer 类中输入的 K-V 对类型（　　）。
　　A. 分离　　　　B. 不一致　　　C. 合并　　　D. 一致
（4）MapReduce 中常见的排序方式有（　　）。
　　A. 部分排序　　B. 全排序　　　C. 选择排序　D. 二次排序
（5）MapReduce 提供的采样器有（　　）。
　　A. 随机采样器　B. 间隔采样器　C. 切片采样器　D. 分类采样器

3. 思考题

（1）MapReduce 的特点和作用是什么？
（2）MapReduce 中解决数据倾斜问题的方法主要有哪些？

4. 编程题

求给定日期的最高温度。

待处理数据内容：201701082.6、201701066、2017020810、2017030816.33、2017060833.0，前 8 位是日期，其后是温度。

第5章 ZooKeeper分布式协调服务

本章学习目标
- 理解 ZooKeeper 的工作原理
- 熟悉 ZooKeeper 的安装
- 掌握 ZooKeeper 的客户端编程方法

ZooKeeper 是 Hadoop 集群管理中必不可少的组件，提供了一套分布式集群管理机制。在 ZooKeeper 的协调下，Hadoop 集群可以实现高可用，保证了集群的稳定性，对于实际生产环境来说意义重大。

5.1 认识 ZooKeeper

5.1.1 ZooKeeper 简介

ZooKeeper 是开源的分布式应用程序协调服务。ZooKeeper 提供了同步服务、命名服务、组服务、配置管理服务，较好地解决了 Hadoop 中经常出现的死锁、竞态条件等问题。

死锁是在执行两个或两个以上的进程时，由竞争资源或彼此通信造成的阻塞现象。竞态条件是指在执行两个或两个以上的进程时，进程执行顺序对执行后的结果存在影响。

ZooKeeper 可以与需要保证高可用的 Hadoop 组件搭配使用，如 HA 模式下的 HDFS、HA 模式下的 YARN、HBase。

5.1.2 ZooKeeper 的设计目的

ZooKeeper 提供一个协调方便、易于编程的环境，能够减轻分布式应用程序所承担的协调任务，其设计目的主要体现在以下几个方面。

（1）一致性。客户不论连接到哪个服务器，看到的都是相同的视图。

（2）实时性。ZooKeeper 的数据存放在内存当中，可以做到高吞吐、低延迟。

（3）可靠性。组成 ZooKeeper 服务的服务器必须互相知道其他服务器的存在。

（4）有序性。例如，ZooKeeper 为每次更新操作赋予一个版本号，此版本号是唯一、有序的。

（5）原子性。ZooKeeper 的客户端在读取数据时，只有成功或失败两种状态，不会出现只读取部分数据的情况。

5.1.3 ZooKeeper 的系统模型

ZooKeeper 的系统模型如图 5.1 所示。

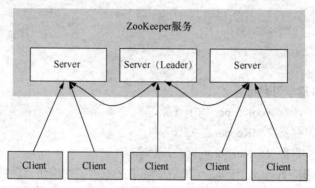

图 5.1　ZooKeeper 的系统模型

ZooKeeper 的系统模型包含服务器（Server）和客户端（Client）。

（1）客户端可以连接到 ZooKeeper 集群的任意服务器。客户端与服务器通过 TCP 建立连接，主要负责发送请求和心跳消息，获取响应和监视事件。

TCP（Transmission Control Protocol，传输控制协议）是一种面向连接的、可靠的、基于字节流的传输层通信协议。

（2）如果客户端与服务器的 TCP 连接中断，客户端自动尝试连接到其他服务器。

（3）客户端第一次连接到一台服务器时，该服务器会为客户端建立一个会话。当该客户端连接到其他的服务器时，新的服务器会为客户端重新建立一个会话。

5.1.4 ZooKeeper 中的角色

ZooKeeper 中主要有三种角色：Leader（领导者）、Follower（跟随者）和 Observer（观察者）。

1. Leader

Leader 是 ZooKeeper 集群中的领导者，集群中有且只有一个，主要负责以下工作。

（1）调度和处理事务请求，保证事务按顺序进行处理。

（2）调度 ZooKeeper 集群内部各个服务器。

2. Follower

Follower 是 ZooKeeper 集群中的跟随者，集群中有多个，主要负责以下工作。

（1）处理客户端非事务请求，转发事务请求给 Leader 服务器。

（2）参与 Leader 选举投票。

3. Observer

Observer 和 Follower 类似，负责处理客户端非事务请求，转发事务请求给 Leader 服务器。Observer 与 Follower 的区别在于：Observer 不参与 Leader 选举投票。

Observer 是从 ZooKeeper 3.3.0 版本开始新增的角色。当 ZooKeeper 集群节点增多时，为了支持更多的客户端,需要增加更多服务器,然而服务器增多后,投票阶段耗费时间过多,影响集群性能。为了增强集群的扩展性,保证数据的高吞吐量,引入 Observer。

5.1.5　ZooKeeper 的工作原理

ZooKeeper 的核心工作原理是:选举一台服务器作为 Leader 对外提供服务,当该 Leader 出问题时,从剩下的多台服务器中快速选出一个新 Leader 继续对外提供服务。

ZooKeeper 实现分布式协调服务,其本身是以集群的形式存在的,采用 ZAB(ZooKeeper Atomic Broadcast,ZooKeeper 原子广播)算法保证自身的高可用性。

ZAB 算法包含两个无限重复的阶段:领导选举(Leader Election)和原子广播(Atomic Broadcast)。

1. 领导选举

领导选举是 ZooKeeper 集群中的多台服务器选举一台服务器作为 Leader 的过程。Leader 之外的服务器为 Follower。

ZooKeeper 规定集群中半数以上的服务器保持正常运行,集群才能对外提供服务。如果 ZooKeeper 集群有 7 台服务器,要保证集群能对外提供服务,最多允许 3 台服务器出现故障;如果 ZooKeeper 集群有 8 台服务器,要保证集群能对外提供服务,也是最多允许 3 台服务器出现故障。安装 ZooKeeper 集群时,一般设置服务器的数量为奇数。

Leader 被选举出来后,如果 ZooKeeper 集群中半数以上的服务器保持正常运行,则选举阶段结束,进入原子广播阶段。

要进行 Leader 选举,至少需要两台机器。以三台机器组成的服务器集群为例,在集群初始化阶段,当服务器 Server1 启动时,无法进行和完成 Leader 选举,当第二台服务器 Server2 启动时,两台机器可以相互通信,每台机器都试图找到 Leader,于是进入 Leader 选举阶段。选举过程如下。

(1)每台服务器发出一个投票。初始情况下,Server1 和 Server2 都会推举自己作为 Leader。投票信息包含所推举的服务器的 myid(机器编号)和 zxid(事务编号),用(myid,zxid)表示,此时 Server1 的投票为(1,0),Server2 的投票为(2,0),它们各自将投票发给集群中其他机器。

(2)接收来自各台服务器的投票。集群的每台服务器收到投票后,先判断该投票的有效性,如检查是否为本轮投票,是否来自 LOOKING(搜寻)状态的服务器。

(3)处理投票。针对每一次投票,服务器都需要将别人的投票和自己的投票进行 PK,PK 规则如下。优先检查 zxid,zxid 比较大的服务器优先作为 Leader。如果 zxid 相同,那么就比较 myid,myid 较大的服务器作为 Leader。对于 Server1 而言,它的投票是(1,0),接收到 Server2 的投票为(2,0),首先比较两者的 zxid,均为 0,然后比较 myid,此时 Server2 的 myid 较大,于是 Server1 更新自己的投票为(2,0),重新投票。对于 Server2 而言,自己的投票无须更新,它只需要再次向集群中所有机器发出上一次的投票信息。

(4)统计投票。每次投票后,服务器都会统计投票情况,判断是否有过半机器接收到相同的投票信息。对于 Server1、Server2 而言,一旦统计出集群中有两台机器接收到了投票信息(2,0),便认为已经选出了 Leader。

(5)改变服务器状态。一旦确定了 Leader,每台服务器就会更新自己的状态,Follower 变更状态为 FOLLOWING(跟随),Leader 变更状态为 LEADING(领导)。

2. 原子广播

原子广播阶段首先需要同步 Leader 和 Follower 的数据。当 Leader 出现故障或者 Leader 失去大多数的 Follower 时，ZooKeeper 集群会在 Follower 中快速选举一个 Leader，保证客户端无论连接哪个 ZooKeeper 服务器，获取的数据都是一致的。保证数据一致性的过程称为原子广播。

5.2 ZooKeeper 安装和常用命令

本书使用的是 3.4.10 版本的 ZooKeeper，安装包见附录。用 Java 编写的 ZooKeeper 运行在虚拟机上，需要提前安装 JDK 并配置好 Java 环境。本书主要讲述 ZooKeeper 的两种安装模式：单机模式和全分布式。

5.2.1 ZooKeeper 单机模式

ZooKeeper 单机模式，安装步骤如下。

（1）将 ZooKeeper 安装包 zookeeper-3.4.10 放到虚拟机 qf01 的 /root/Downloads/ 目录下，切换到 root 用户，新建目录 /mysoft，解压 ZooKeeper 安装包到 /mysoft 目录下。

```
[root@qf01 ~]# mkdir /mysoft
[root@qf01 ~]# tar -zxvf /root/Downloads/zookeeper-3.4.10.tar.gz -C /mysoft/
```

（2）切换到 /mysoft 目录下，将 zookeeper-3.4.10 重命名为 zookeeper。

```
[root@qf01 ~]# cd /mysoft/
[root@qf01 mysoft]# mv zookeeper-3.4.10 zookeeper
```

（3）打开 /etc/profile 文件，配置 ZooKeeper 环境变量。

```
[root@qf01 mysoft]# vi /etc/profile
```

在文件末尾添加如下三行内容。

```
# ZooKeeper environment variables
export ZOOKEEPER_HOME=/mysoft/zookeeper
export PATH=$PATH:$ZOOKEEPER_HOME/bin
```

（4）使环境变量生效。

```
[root@qf01 mysoft]# source /etc/profile
```

（5）将文件 /mysoft/zookeeper/conf/zoo_sample.cfg 重命名为 zoo.cfg（ZooKeeper 的配置文件）。

```
[root@qf01 mysoft]# cd /mysoft/zookeeper/conf/
[root@qf01 conf]# mv zoo_sample.cfg zoo.cfg
```

（6）启动 ZooKeeper 的服务器。

```
[root@qf01 conf]# zkServer.sh start
ZooKeeper JMX enabled by default
Using config: /mysoft/zookeeper/bin/../conf/zoo.cfg
Starting zookeeper ... STARTED
```

（7）检测 ZooKeeper 服务器是否启动成功有两种方法。
① 查看 ZooKeeper 服务器的启动状态。

```
[root@qf01 conf]# zkServer.sh status
```

出现如下内容表明 ZooKeeper 服务器启动成功。

```
ZooKeeper JMX enabled by default
Using config: /mysoft/zookeeper/bin/../conf/zoo.cfg
Mode: standalone
```

② 用 jps 命令查看 ZooKeeper 服务器的 QuorumPeerMain 进程是否启动。

```
[root@qf01 conf]# jps
11716 Jps
10412 QuorumPeerMain
```

出现 QuorumPeerMain 进程表明 ZooKeeper 服务器启动成功。QuorumPeerMain 是 ZooKeeper 集群的启动入口。

（8）关闭 ZooKeeper 服务器。

```
[root@qf01 conf]# zkServer.sh stop
ZooKeeper JMX enabled by default
Using config: /mysoft/zookeeper/bin/../conf/zoo.cfg
Stopping zookeeper ... STOPPED
```

5.2.2 ZooKeeper 全分布式

1. 安装步骤

ZooKeeper 全分布式，又称 ZooKeeper 集群模式，安装步骤如下。

（1）修改 ZooKeeper 的配置文件 zoo.cfg。

```
[root@qf01 conf]# vi /mysoft/zookeeper/conf/zoo.cfg
```

将 dataDir=/tmp/zookeeper 修改为如下内容。

```
dataDir=/mysoft/zookeeper/zkdata
```

在文件末尾添加如下三行内容。

```
server.1=qf01:2888:3888
server.2=qf02:2888:3888
server.3=qf03:2888:3888
```

其中，"1" "2" "3" 是 myid，要求是 1~255 的整数；qf01、qf02、qf03 代表其对应的主机地址。2888 是 Leader 端口，负责和 Follower 进行通信。3888 是 Follower 端口，负责推选 Leader。

（2）新建目录/mysoft/zookeeper/zkdata，在该目录下新建文件 myid。

```
[root@qf01 conf]# mkdir /mysoft/zookeeper/zkdata
[root@qf01 conf]# vi /mysoft/zookeeper/zkdata/myid
```

在 myid 文件中填写如下内容。

```
1
```

（3）将/soft/zookeeper/分发到虚拟机 qf02、qf03。

```
[root@qf01 conf]# scp -r /mysoft/zookeeper/ root@qf02:/mysoft/zookeeper
[root@qf01 conf]# scp -r /mysoft/zookeeper/ root@qf03:/mysoft/zookeeper
```

（4）修改虚拟机 qf02 的/mysoft/zookeeper/zkdata/myid 文件。

```
[root@qf02 ~]# vi /mysoft/zookeeper/zkdata/myid
```

将 myid 文件中的内容替换为如下内容。

```
2
```

（5）修改虚拟机 qf03 的 /mysoft/zookeeper/zkdata/myid 文件。

```
[root@qf03 ~]# vi /mysoft/zookeeper/zkdata/myid
```

将 myid 文件中的内容替换为如下内容。

```
3
```

（6）将虚拟机 qf01 的系统环境变量分发到虚拟机 qf02、qf03。

```
[root@qf01 conf]# scp /etc/profile qf02:/etc/
[root@qf01 conf]# scp /etc/profile qf03:/etc/
```

（7）分别使虚拟机 qf02、qf03 的环境变量生效。

```
[root@qf02 ~]# source /etc/profile
[root@qf03 ~]# source /etc/profile
```

（8）分别启动虚拟机 qf01、qf02、qf03 的 ZooKeeper 服务器。

```
[root@qf01 conf]# zkServer.sh start
[root@qf02 ~]# zkServer.sh start
[root@qf03 ~]# zkServer.sh start
```

（9）分别查看各虚拟机的 ZooKeeper 服务器启动状态。

查看虚拟机 qf01 的 ZooKeeper 服务器启动状态。

```
[root@qf01 conf]# zkServer.sh status
ZooKeeper JMX enabled by default
Using config: /mysoft/zookeeper/bin/../conf/zoo.cfg
Mode: follower
```

查看虚拟机 qf02 的 ZooKeeper 服务器启动状态。

```
[root@qf02 ~]# zkServer.sh status
ZooKeeper JMX enabled by default
Using config: /mysoft/zookeeper/bin/../conf/zoo.cfg
Mode: leader
```

查看虚拟机 qf03 的 ZooKeeper 服务器启动状态。

```
[root@qf03 ~]# zkServer.sh status
ZooKeeper JMX enabled by default
Using config: /mysoft/zookeeper/bin/../conf/zoo.cfg
Mode: follower
```

查看启动状态返回的结果中，出现 Mode: follower 或 Mode: leader，表明 ZooKeeper 服务器启动成功。

2. 启动和关闭 ZooKeeper 集群

启动和关闭 ZooKeeper 集群需要在每台虚拟机上启动和关闭 ZooKeeper 服务器。在实际工作中，使用的服务器可能会比较多，在每台服务器上进行重复的操作效率不高。为了方便地启动和关闭 ZooKeeper 集群，可以编写启动脚本 xzk.sh，具体步骤如下。

（1）在虚拟机 qf01 的 /usr/local/bin/ 目录下，新建 xzk.sh 脚本文件，xzk.sh 文件内容如下。

```bash
#!/bin/bash
cmd=$1
if [ $# -gt 1 ] ; then echo param must be 1 ; exit ; fi
for (( i=1 ; i<=3 ; i++ )) ; do
   tput setaf 5
   echo ============ qf0$i $@ ============
   tput setaf 9
   ssh qf0$i "source /etc/profile ; zkServer.sh $cmd"
done
```

（2）为 xzk.sh 脚本拥有者添加执行权限。

```
[root@qf01 bin]# chmod u+x xzk.sh
```

（3）通过 xzk.sh 脚本的 start 和 stop 命令，在虚拟机 qf01 上同时启动和关闭虚拟机 qf01、qf02、qf03 的 ZooKeeper 服务器。

```
[root@qf01 bin]# xzk.sh start
[root@qf01 bin]# xzk.sh stop
```

至此，ZooKeeper 全分布式安装完成。

5.2.3 ZooKeeper 服务器常用脚本

ZooKeeper 服务器的常用脚本在 /mysoft/zookeeper/bin/ 目录下，具体含义如表 5.1 所示。

表 5.1　　　　　　　　　　　ZooKeeper 服务器的常用脚本

脚本	含义
zkCleanup.sh	清理 ZooKeeper 历史数据，包括事务日志文件和快照数据文件
zkCli.sh	开启一个简易的 ZooKeeper 客户端
zkEnv.sh	设置 ZooKeeper 的环境变量
zkServer.sh	ZooKeeper 服务器的启动、关闭和重启脚本

5.2.4 ZooKeeper 客户端节点和命令

1. ZooKeeper 客户端的节点

（1）ZooKeeper 客户端的文件系统。

ZooKeeper 客户端的文件系统使用了树形目录结构，与 Linux 文件系统类似。ZooKeeper 客户端的目录项都被称为节点（Znode）。每个节点拥有数据、数据长度、创建时间、修改时间、子节点数量等信息。

注意：此处的节点要区别于 NameNode、DataNode 等。

（2）ZooKeeper 节点的 3 种类型。

- 永久节点：在 ZooKeeper 客户端退出后，不会自动删除的节点。ZooKeeper 客户端默认创建永久节点。
- 临时节点：在 ZooKeeper 客户端退出后，会自动删除的节点。临时节点不能有子节点。使用者可以通过临时节点判断分布式服务的打开或关闭。
- 顺序节点：节点名称末尾会自动附加一个 10 位序列号的节点。

（3）ZooKeeper 节点的权限控制。

在实际生产环境中，多个应用常使用相同的 ZooKeeper，但是不同应用系统很少使用共同的数据。

鉴于这种情况，ZooKeeper 采用 ACL（Access Control List，访问控制列表）策略来进行权限控制，类似于 Linux 文件系统的权限控制。ZooKeeper 的节点定义了 5 种权限。

- CREATE：创建子节点的权限。
- READ：获取子节点数据和子节点列表的权限。
- WRITE：更新节点数据的权限。
- DELETE：删除子节点的权限。
- ADMIN：设置节点 ACL 的权限。

2. 打开 ZooKeeper 客户端

以下操作在 ZooKeeper 全分布式下进行。

（1）打开 ZooKeeper 本地客户端。

在虚拟机 qf01、qf02、qf03 的 ZooKeeper 服务器都启动的情况下，在虚拟机 qf01 中使用如下命令打开客户端。

```
[root@qf01 conf]# zkCli.sh
```

出现如下内容，表明 ZooKeeper 客户端启动成功。

```
[zk: localhost:2181(CONNECTED) 0]
```

退出 ZooKeeper 客户端的命令如下。

```
[zk: localhost:2181(CONNECTED) 0] quit
```

（2）打开指定的 ZooKeeper 远程客户端。

例如，在虚拟机 qf01 上打开远程虚拟机 qf02 的 ZooKeeper 客户端，可使用如下命令。

```
zkCli.sh -server qf02:2181
```

其中，qf02 代表虚拟机 qf02 的主机地址。

```
[root@qf01 conf]# zkCli.sh -server qf02:2181
```

出现如下内容，表明虚拟机 qf02 的 ZooKeeper 客户端启动成功。

```
[zk: qf02:2181(CONNECTED) 0]
```

3. ZooKeeper 客户端的命令

（1）使用 help 命令查看 ZooKeeper 客户端的命令。

```
[zk: qf02:2181(CONNECTED) 0] help
ZooKeeper -server host:port cmd args
        stat path [watch]
        set path data [version]
        ls path [watch]
        delquota [-n|-b] path
        ls2 path [watch]
        setAcl path acl
        setquota -n|-b val path
        history
        redo cmdno
        printwatches on|off
        delete path [version]
        sync path
        listquota path
```

```
rmr path
get path [watch]
create [-s] [-e] path data acl
addauth scheme auth
quit
getAcl path
close
connect host:port
```

（2）ZooKeeper 客户端的常用命令。

ZooKeeper 客户端的常用命令及具体含义，如表 5.2 所示。

表 5.2　　　　　　　　　　　　ZooKeeper 客户端的常用命令

命令	含义
ls /	查看指定路径下的节点
create /a Tom	创建永久节点/a 并添加数据 Tom（create 命令无法实现递归创建）
create -e /b Tom	创建临时节点/b 并添加数据 Tom
create -s /c Tom	创建顺序节点/c 并添加数据 Tom
get /a	查看/a 节点的数据和状态（get 命令后必须用绝对路径）
stat /a	查看/a 节点的状态（和 get 的区别在于不显示数据）
set /a Jack	将/a 节点数据 Tom 修改为 Jack
delete /a	删除节点/a（不能递归删除）
rmr /a	递归删除节点/a
connect qf01:2181	连接虚拟机 qf01 的 ZooKeeper 客户端
quit	退出 ZooKeeper 客户端

注意：ZooKeeper 客户端中操作节点的命令必须使用绝对路径。

5.3 ZooKeeper 客户端编程

ZooKeeper 提供了 Java API 来方便开发者进行客户端编程，并根据项目需求操作服务器上的数据。

5.3.1 配置开发环境

在 IntelliJ IDEA 中项目 testHadoop 下新建模块 testZK，修改 pom.xml 文件内容如下。

```xml
<?xml version="1.0" encoding="UTF-8"?>
<project xmlns="http://maven.apache.org/POM/4.0.0"
     xmlns:xsi="http://www.w3.org/2001/XMLSchema-instance"
     xsi:schemaLocation="http://maven.apache.org/POM/4.0.0
     http://maven.apache.org/xsd/maven-4.0.0.xsd">
  <modelVersion>4.0.0</modelVersion>
  <groupId>com.qf</groupId>
  <artifactId>testZK</artifactId>
  <version>1.0-SNAPSHOT</version>
  <dependencies>
    <dependency>
      <groupId>org.apache.zookeeper</groupId>
      <artifactId>zookeeper</artifactId>
      <version>3.4.10</version>
```

```xml
            </dependency>
            <dependency>
                <groupId>junit</groupId>
                <artifactId>junit</artifactId>
                <version>4.12</version>
            </dependency>
        </dependencies>
</project>
```

5.3.2 Java 程序操作 ZooKeeper 客户端

编写 Java 程序实现对 ZooKeeper 客户端的操作。

【例 5-1】OperateZK.java

```
1   public class OperateZK {
2       @Test
3       /**
4        * 1.列出根节点下的子节点
5        */
6       public void testList() throws Exception {
7           //定义连接串，以逗号分隔ip:port，客户端端口2181
8           String conn = "192.168.142.131:2181,192.168.142.132:2181,192.168.142.133:2181";
9           //参数1是连接串，参数2是超时时长，参数3是观察者（null）
10          ZooKeeper zk = new ZooKeeper(conn, 5000, null);
11          //列出根目录下的子节点
12          List<String> children = zk.getChildren("/", false);
13          for (String child : children) {
14              System.out.println(child);
15          }
16      }
17      @Test
18      /**
19       * 2.创建节点/a，写入数据Tom
20       */
21      public void testCreate() throws Exception {
22          String conn = "192.168.142.131:2181,192.168.142.132:2181,192.168.142.133:2181";
23          ZooKeeper zk = new ZooKeeper(conn, 5000, null);
24          //创建节点/a，写入数据Tom
25          String s = zk.create("/a", "Tom".getBytes(), ZooDefs.Ids.OPEN_ACL_UNSAFE, CreateMode.PERSISTENT);
26          System.out.println(s);
27          zk.close();
28      }
29      @Test
30      /**
31       * 3.获取指定节点/a的数据
32       */
33      public void testGet() throws Exception {
34          String conn = "192.168.142.131:2181,192.168.142.132:2181,192.168.142.133:2181";
35          ZooKeeper zk = new ZooKeeper(conn, 5000, null);
```

```
36              //测试 stat 在获取数据时数据会被写入
37              Stat stat = new Stat();
38              System.out.println(stat.toString());
39              //获取根节点数据
40              byte[] a = zk.getData("/a", false, stat);
41              System.out.println(stat.toString());
42              System.out.println(new String(a));
43              zk.close();
44          }
45      }
```

运行结果如下。

```
ZooKeeper
/a
Tom
```

5.4 ZooKeeper 典型应用场景

5.4.1 数据发布与订阅

数据发布与订阅模型，即配置中心，顾名思义就是发布者将数据发布到 ZooKeeper 节点上，供订阅者动态获取数据，实现配置信息的集中式管理和动态更新。全局配置信息、服务式框架的服务地址列表等非常适合使用该模型。

（1）把应用中用到的一些配置信息放到 ZooKeeper 上进行集中管理。这类场景通常是这样：应用在启动时会主动来获取一次配置，同时在节点上注册一个 Watcher（监听者），这样一来，以后每次配置有更新的时候，客户端订阅者都会收到通知。

（2）分布式搜索服务中，索引的元信息和服务器集群机器的节点状态存放在 ZooKeeper 的一些指定节点，供客户端订阅使用。

（3）分布式日志收集系统。这个系统的核心工作是收集分布在不同机器上的日志。收集器通常按照应用来分配收集任务单元，因此需要在 ZooKeeper 上创建一个以应用名作为 Path（路径）的节点 P，并将这个应用的所有机器 IP 地址，以子节点的形式注册到节点 P 上，这样一来，机器变动的时候就能够实时通知收集器调整任务分配。

（4）系统中有些信息需要动态获取，并且存在手动修改信息的询问。这时通常要暴露出接口，如 JMX（Java Management Extensions，Java 管理扩展）接口，来获取一些运行时的信息。引入 ZooKeeper 之后，开发者就不用自己去实现一套方案了，只要将这些信息存放到指定的 ZooKeeper 节点上即可。

注意：上面提到的应用场景的默认前提是数据量很小而数据更新比较快。

5.4.2 命名服务

命名服务也是分布式系统中比较常见的一类场景。在分布式系统中，通过使用命名服务，客户端应用能够根据指定的名字来获取资源或服务的地址、提供者等信息。被命名的实体可以是集群中的机器、提供的服务地址、远程对象等。其中较为常见的是一些分布式服务框架中的服务地址列表。通过调用 ZooKeeper 提供的创建节点的 API，很容易创建一个全局唯一的 Path，这个 Path 就可以作为一个名字。

阿里巴巴集团的开源分布式服务框架 Dubbo 使用 ZooKeeper 来提供其命名服务，维护全局的服务地址列表。

在 Dubbo 实现中：服务提供者在启动时，向 ZooKeeper 上的指定节点/dubbo/${serviceName}/providers 目录下写入自己的 URL 地址，完成服务的发布；服务消费者在启动时，订阅/dubbo/${serviceName}/providers 目录下的提供者 URL 地址，并向/dubbo/${serviceName}/consumers 目录下写入自己的 URL 地址。

注意：所有向 ZooKeeper 注册的地址都是临时节点，这样能够保证服务提供者和消费者自动感应资源的变化。另外，Dubbo 还有针对服务粒度的监控，监控方法是订阅/dubbo/${serviceName}目录下所有服务提供者和服务消费者的信息。

5.4.3 分布式锁

分布式锁，主要得益于 ZooKeeper 保证了数据的强一致性。锁服务可以分为两类：一个是保持独占，另一个是控制时序。

保持独占，就是所有试图获取这把锁的客户端，最终只有一个可以成功，从而执行相应的操作（通常的做法是把 ZooKeeper 上的一个节点看成一把锁，通过创建临时节点的方式来实现）。

控制时序，就是所有试图获取这把锁的客户端，最终都能够获取，只是有个全局时序。其做法和上面基本类似，只是这里 /distribute_lock 已经预先存在，客户端在它下面创建临时顺序节点。ZooKeeper 的父节点（/distribute_lock）保证子节点创建的时序性，从而形成每个客户端的全局时序。

5.5 本章小结

本章主要对 ZooKeeper 的基本原理、安装、常用命令、客户端编程、典型应用场景进行了讲解，应重点理解 ZooKeeper 的 Leader 选举机制。

5.6 习题

1. 填空题

（1）ZooKeeper 提供了_____、命名服务、组服务、配置管理等服务，较好地解决了 Hadoop 中经常出现的死锁、竞态条件等问题。

（2）ZooKeeper 可以与需要保证高可用的 Hadoop 组件搭配使用，例如，_____、HA 模式下的 YARN、HBase。

（3）ZooKeeper 中主要有三种角色_____、Follower 和 Observer。

（4）ZooKeeper 的节点主要有_____、临时节点、顺序节点 3 种类型。

（5）ZAB 算法包含两个无限重复的阶段：_____和原子广播。

2. 选择题

（1）ZooKeeper 启动时最多会监听（　　）个端口。

  A．1   B．2   C．3   D．4

（2）下列（　　）操作可以设置一个监听 Watcher。

　　　A. getData　　　B. GetChildren　　　C. exists　　　D. setData

（3）ZooKeeper 的系统模型分为服务器和（　　）。

　　　A. 计算模型　　　B. 用户界面　　　C. 客户端　　　D. 存储器

（4）ZooKeeper 服务器的常用脚本有（　　）。

　　　A. zkCleanup.sh　　　B. zkEnv.sh　　　C. zkCli.sh　　　D. 以上都是

（5）（　　）不是 ZooKeeper 的设计目的。

　　　A. 最终一致性　　　B. 实时性　　　C. 可靠性　　　D. 循环性

3. 思考题

（1）ZooKeeper 的核心工作原理是什么？

（2）ZooKeeper 的一致性指什么？

第6章 Hadoop 2.0新特性

本章学习目标
- 熟悉 Hadoop 2.0 的改进与提升
- 理解 YARN 架构的原理
- 理解 Hadoop 的 HA 模式

Hadoop 诞生以来，主要分为 Hadoop 1.0、Hadoop 2.0、Hadoop 3.0 三个系列的多个版本。目前最常见的是 Hadoop 2.0 系列。Hadoop 2.0 指的是第 2 代 Hadoop，它是从 Hadoop 1.0 发展而来的，相对于 Hadoop 1.0 有很多改进。本章对 Hadoop 2.0 新特性进行详细讲解。

6.1 Hadoop 2.0 的改进

Hadoop 1.0 由 MapReduce 和 HDFS 组成，在高可用、扩展性方面存在一些问题。Hadoop 2.0 由 HDFS、MapReduce 和 YARN 三个分支构成。Hadoop 1.0 和 Hadoop 2.0 的区别如表 6.1 所示。

表 6.1 Hadoop 1.0 和 Hadoop 2.0 的区别

组件	Hadoop 1.0 的局限和不足	Hadoop 2.0 的改进
HDFS	NameNode 存在单点故障风险	HDFS 引入了高可用机制
MapReduce	JobTracker 存在断电故障风险，且内存扩展受限	引入了一个资源管理调度框架 YARN

6.1.1 HDFS 存在的问题

（1）NameNode 单点故障使其难以应用于在线场景。
（2）NameNode 压力过大，且内存受限，影响系统扩展性。

6.1.2 MapReduce 存在的问题

（1）JobTracker 单点故障。
（2）JobTracker 访问压力大，影响系统扩展性。
（3）难以支持除 MapReduce 之外的计算框架，如 Spark、Storm、Tez 等。

6.1.3 HDFS 2.0 解决 HDFS 1.0 中的问题

（1）解决单点故障。HDFS HA 有两个 NameNode，如果活跃 NameNode 发生故障，则切换到备用 NameNode。

（2）解决内存受限问题。HDFS Federation（联邦）水平扩展方案支持多个 NameNode，每个 NameNode 分管一部分目录，所有 NameNode 共享所有 DataNode 存储资源。

（3）仅是架构上发生了变化，使用方式不变，对 HDFS 使用者透明。

6.2 YARN 资源管理框架

6.2.1 YARN 简介

YARN 是基于 MapReduce 的作业调度与集群资源管理框架。YARN 主要由 ResourceManager（资源管理器）、NodeManager（节点管理器）、ApplicationMaster（应用程序管理器）和 Container（容器）四个组件构成。

YARN 允许开发者根据需求在集群中运行不同版本的 MapReduce，获取更为便捷的服务。

6.2.2 YARN 架构

YARN 总体上是主/从（Master/Slave）结构，在整个资源管理框架中，ResourceManager 为 Master，NodeManager 为 Slave。ResourceManager 负责对各个 NodeManager 上的资源进行统一管理和调度。用户提交一个应用程序时，需要提供一个用以跟踪和管理这个程序的 ApplicationMaster。ApplicationMaster 负责向 ResourceManager 申请资源。不同的 ApplicationMaster 被分布到不同的节点上，它们之间不会相互影响。YARN 架构如图 6.1 所示。

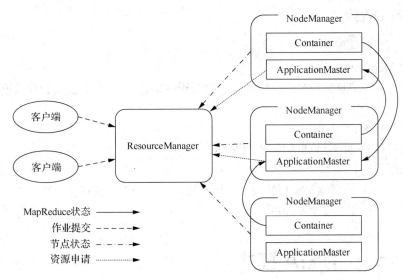

图 6.1 YARN 架构

1. YARN 主要组件的作用

YARN 架构中 4 个主要组件的作用如下。

（1）ResourceManager 负责接收客户端的任务请求，接收和监控 NodeManager 的资源分配与调度，启动和监控 ApplicationMaster。

（2）NodeManager 负责节点上的资源管理，启动 Container 运行任务，上报节点资源使用情况、Container 运行情况给 ApplicationMaster。

（3）ApplicationMaster 负责单个应用作业的任务管理和调度，向 ResourceManager 申请资源，向 NodeManager 发出启动 Container 的指令，接收 NodeManager 的任务处理状态信息。

（4）Container 是 YARN 中的资源抽象，主要负责对任务运行环境的抽象，描述一系列信息：任务运行资源（节点、内存、CPU）、任务启动命令和任务运行环境。YARN 会为每个任务分配一个 Container，且每个任务只能使用分配的 Container 中描述的资源。

2．YARN 的工作流程

（1）客户端向 ResourceManager 提交程序。

（2）ResourceManager 向 NodeManager 分配一个 Container，并在该 Container 中启动 ApplicationMaster。

（3）ApplicationMaster 向 ResourceManager 注册，这样客户端可以直接通过 ResourceManager 查看应用程序的运行状态，然后 ApplicationMaster 将为各个任务申请资源，并监控任务的运行状态，直到运行结束。

（4）ApplicationMaster 采用轮询的方式通过 RPC 协议向 ResourceManager 申请和领取资源，异步完成资源的协调。

（5）ApplicationMaster 申请到资源后，与对应的 NodeManager 通信，要求它启动任务。

（6）NodeManager 为任务设置好运行环境（环境变量、jar 包、二进制程序等）后，将任务启动命令写到一个脚本中，并通过运行该脚本启动任务。

（7）各个任务通过某个 RPC 协议向 ApplicationMaster 汇报自己的状态和进度，让 ApplicationMaster 随时掌握各个任务的运行状态，以便在任务失败时重新启动任务。

（8）应用程序运行完成后，注销并关闭 ApplicationMaster。

6.2.3 YARN 的优势

YARN 的优势主要体现在以下几个方面。

（1）更快的 MapReduce 计算。

YARN 利用异步模型对 MapReduce 框架的关键逻辑结构进行了重写，使其计算速度比 MapReduce 1.0 更快。

（2）支持多种框架。

用户可以将各种各样的计算框架移植到 YARN 上。

（3）框架升级容易。

在 YARN 中，各种计算框架被封装成一个用户程序库（lib）存放在客户端，当需要对计算框架进行升级时，只要升级用户程序库即可。

6.3 Hadoop 的 HA 模式

6.3.1 HA 模式简介

Hadoop 的 HA 模式是在 Hadoop 全分布式基础上，利用 ZooKeeper 等协调工具配置的高可用的

Hadoop 集群模式。

Hadoop 的 HA 模式搭建规划如表 6.2 所示。本书中 Hadoop 的 HA 模式是通过 ZooKeeper 来实现的，因此需要在 Hadoop 的配置文件里对 ZooKeeper 进行相关设置。

表 6.2　　　　　　　　　　　　　　Hadoop 的 HA 模式搭建规划

主机名	IP 地址	相关进程
qf01	192.168.142.131	NameNode、DataNode、DFSZKFailoverController、QuorumPeerMain、JournalNode、ResourceManager、NodeManager
qf02	192.168.142.132	NameNode、DataNode、DFSZKFailoverController、QuorumPeerMain、JournalNode、ResourceManager、NodeManager
qf03	192.168.142.133	DataNode、NodeManager、QuorumPeerMain、JournalNode

Hadoop 的 HA 模式包括两部分：HDFS 的 HA 模式、YARN 的 HA 模式。

6.3.2　HDFS 的 HA 模式

HDFS 的 HA 模式下有两个 NameNode，一个 NameNode 处于活跃（Active）状态，另一个 NameNode 处于备用（Standby）状态。活跃的 NameNode 负责 Hadoop 集群中的所有客户端操作，而备用的 NameNode 只是充当从属服务器，维持一定的状态以便在必要时进行快速故障转移。

Hadoop 的 HA 模式出现的原因是 Hadoop 集群存在单点故障问题。单点故障是指对于只有一个 NameNode 的 Hadoop 集群，NameNode 出现故障会导致 Hadoop 集群无法正常工作。HA 模式下的 Hadoop 集群解决了单点故障问题，其基本原理是：当一个 NameNode 出现故障时，HA 模式下的 Hadoop 集群通过 ZooKeeper 等协调工具快速启动备用的 NameNode，确保 Hadoop 集群的高可用性。

1. HDFS 的 HA 模式搭建

由于 HDFS 的 HA 模式搭建步骤较多，建议在搭建之前先对虚拟机 qf01、qf02、qf03 拍摄快照。这样如果搭建过程中系统报错，就可以快速恢复到搭建以前的状态。

本书中 HDFS 的 HA 模式搭建是在前面的 Hadoop 集群和 ZooKeeper 集群基础上进行的，具体步骤如下。

（1）确保 Hadoop 集群和 ZooKeeper 集群处于关闭状态。

（2）修改虚拟机 qf01 的 /usr/local/hadoop-2.7.3/etc/hadoop/core-site.xml 文件。

```
[root@qf01 ~]# vi /usr/local/hadoop-2.7.3/etc/hadoop/core-site.xml
```

将 core-site.xml 文件中的内容替换如下。

```
<configuration>
    <!-- 指定文件系统的名称-->
    <property>
        <name>fs.defaultFS</name>
        <value>hdfs://qianfeng</value>
    </property>
    <!-- 配置Hadoop运行产生的临时数据存储目录 -->
    <property>
        <name>hadoop.tmp.dir</name>
        <value>/tmp/hadoop-qf01</value>
    </property>
    <!-- 指定自动故障转移的集群 -->
    <property>
```

```xml
        <name>ha.zookeeper.quorum</name>
        <value>qf01:2181,qf02:2181,qf03:2181</value>
    </property>
    <!-- 配置操作 HDFS 的缓存大小 -->
    <property>
        <name>io.file.buffer.size</name>
        <value>4096</value>
    </property>
</configuration>
```

（3）修改虚拟机 qf01 的 /usr/local/hadoop-2.7.3/etc/hadoop/hdfs-site.xml 文件。

```
[root@qf01 ~]# vi /usr/local/hadoop-2.7.3/etc/hadoop/hdfs-site.xml
```

将 hdfs-site.xml 文件中的内容替换如下。

```xml
<configuration>
    <!-- 配置 HDFS 块的副本数（全分布模式默认副本数是 3，最大副本数是 512）-->
    <property>
        <name>dfs.replication</name>
        <value>3</value>
    </property>
    <!-- 设置 HDFS 块的大小 -->
    <property>
        <name>dfs.blocksize</name>
        <value>134217728</value>
    </property>
    <!-- 配置 HDFS 元数据的存储目录 -->
    <property>
        <name>dfs.namenode.name.dir</name>
        <value>/home/hadoopdata/dfs/name</value>
    </property>
    <!-- 配置 HDFS 真正的数据内容（数据块）的存储目录 -->
    <property>
        <name>dfs.datanode.data.dir</name>
        <value>/home/hadoopdata/dfs/data</value>
    </property>
    <!-- 开启通过 Web 操作 HDFS -->
    <property>
        <name>dfs.webhdfs.enabled</name>
        <value>true</value>
    </property>
    <!-- 关闭 HDFS 文件的权限检查 -->
    <property>
        <name>dfs.permissions.enabled</name>
        <value>false</value>
    </property>
    <!-- 配置虚拟服务名 -->
    <property>
        <name>dfs.nameservices</name>
        <value>qianfeng</value>
    </property>
    <!-- 为虚拟服务指定两个 NameNode（目前每个虚拟服务最多可以配置两个 NameNode）-->
    <property>
        <name>dfs.ha.namenodes.qianfeng</name>
```

```xml
        <value>nn1,nn2</value>
    </property>
    <!-- 配置NameNode（nn1）的RPC地址 -->
    <property>
        <name>dfs.namenode.rpc-address.qianfeng.nn1</name>
        <value>qf01:8020</value>
    </property>
    <!-- 配置NameNode（nn2）的RPC地址 -->
    <property>
        <name>dfs.namenode.rpc-address.qianfeng.nn2</name>
        <value>qf02:8020</value>
    </property>
    <!-- 配置NameNode（nn1）的HTTP地址 -->
    <property>
        <name>dfs.namenode.http-address.qianfeng.nn1</name>
        <value>qf01:50070</value>
    </property>
    <!-- 配置NameNode（nn2）的HTTP地址 -->
    <property>
        <name>dfs.namenode.http-address.qianfeng.nn2</name>
        <value>qf02:50070</value>
    </property>
    <!-- 配置JournalNode通信地址 -->
    <property>
        <name>dfs.namenode.shared.edits.dir</name>
        <value>qjournal://qf01:8485;qf02:8485;qf03:8485/qianfeng</value>
    </property>
    <!-- 配置NameNode出现故障时，启用备用NameNode的代理 -->
    <property>
        <name>dfs.client.failover.proxy.provider.qianfeng</name>
        <value>org.apache.hadoop.hdfs.server.namenode.ha.ConfiguredFailoverProxyProvider</value>
    </property>
    <!-- 配置自动故障转移 -->
    <property>
        <name>dfs.ha.automatic-failover.enabled</name>
        <value>true</value>
    </property>
    <!-- 配置防止脑裂的手段，本书使用shell脚本(/bin/true) -->
    <property>
        <name>dfs.ha.fencing.methods</name>
        <value>shell(/bin/true)</value>
    </property>
</configuration>
```

脑裂（split-brain）是指在HA模式下，当联系着的两个节点断开联系时，本来为一个整体的系统，分裂为两个独立节点，这时两个节点开始争抢共享资源，导致系统混乱，数据损坏。

（4）将虚拟机qf01上的Hadoop目录分发到虚拟机qf02、qf03。

```
[root@qf01 ~]# scp -r /usr/local/hadoop-2.7.3 qf02:/usr/local/
[root@qf01 ~]# scp -r /usr/local/hadoop-2.7.3 qf03:/usr/local/
```

（5）在虚拟机qf02上新建SSH公私秘钥对。

```
[root@qf02 ~]# ssh-keygen -t rsa -P '' -f ~/.ssh/id_rsa
```

（6）在虚拟机 qf02 上配置免密登录虚拟机 qf01、qf02、qf03。

```
[root@qf02 ~]# ssh-copy-id root@qf01
[root@qf02 ~]# ssh-copy-id root@qf02
[root@qf02 ~]# ssh-copy-id root@qf03
```

（7）将虚拟机 qf01 上 HDFS 元数据的存储目录（/home/hadoopdata/dfs/name）分发到虚拟机 qf02 的 /home/hadoopdata/dfs/ 目录下。

```
[root@qf01 ~]# scp -r /home/hadoopdata/dfs/name root@qf02:/home/hadoopdata/dfs/name
```

2. 启动 HDFS 的 HA 模式

首次启动 HDFS 的 HA 模式步骤较多，如果首次启动成功，后续启动操作相对简单。

（1）首次启动 HDFS 的 HA 模式，步骤如下。

① 在虚拟机 qf01 上启动 ZooKeeper 集群。

```
[root@qf01 ~]# xzk.sh start
```

② 在虚拟机 qf01 上格式化 ZooKeeper。

```
[root@qf01 ~]# hdfs zkfc -formatZK
```

③ 分别在虚拟机 qf01、qf02、qf03 上启动 JournalNode 进程。

```
[root@qf01 ~]# hadoop-daemon.sh start journalnode
[root@qf02 ~]# hadoop-daemon.sh start journalnode
[root@qf03 ~]# hadoop-daemon.sh start journalnode
```

④ 在虚拟机 qf01 上初始化共享编辑日志。

```
[root@qf01 ~]# hdfs namenode -initializeSharedEdits
```

⑤ 在虚拟机 qf01 上启动 HDFS 进程。

```
[root@qf01 ~]# start-dfs.sh
```

（2）后续启动 HDFS 的 HA 模式，步骤如下。

① 在虚拟机 qf01 上启动 ZooKeeper 集群。

```
[root@qf01 ~]# xzk.sh start
```

② 在虚拟机 qf01 上启动 HDFS 进程。

```
[root@qf01 ~]# start-dfs.sh
```

3. 验证 HDFS 相关进程是否成功启动

在虚拟机 qf01 上查看虚拟机 qf01、qf02、qf03 的 HDFS 相关进程。

```
[root@qf01 ~]# jps
7024 DFSZKFailoverController
6515 NameNode
6633 DataNode
7161 Jps
5210 QuorumPeerMain
5818 JournalNode
[root@qf02~]#jps
```

```
3191 QuorumPeerMain
3736 JournalNode
4362 DataNode
4698 Jps
4573 DFSZKFailoverController
4271 NameNode
[root@qf03~]#jps
3122 QuorumPeerMain
3924 JournalNode
4356 Jps
4153 DataNode
```

出现以上进程，表明 HDFS 相关进程成功启动。

4. 自动故障转移

HDFS 的 HA 模式主要用于实现服务器的自动故障转移，保证 HDFS 的高可用性。而自动故障转移的实现依赖于 ZooKeeper 的如下功能。

（1）故障检测。在 Hadoop 集群中，每个 NameNode 所在的服务器都在 ZooKeeper 中维护一个持久会话。如果当前活跃的 NameNode 服务器出现故障，该服务器在 ZooKeeper 中维持的会话中断，并通知另一个 NameNode 触发故障转移。

（2）指定活跃的 NameNode。如果当前活跃的 NameNode 服务器出现故障，则 ZooKeeper 指定备用的 NameNode 成为活跃的节点。

以上两个功能主要是通过 QuorumPeerMain、JournalNode 和 DFSZKFailoverController（ZKFC）进程实现的。下面对 JournalNode 和 DFSZKFailoverController 进行讲解。

（1）JournalNode 主要用于两个 NameNode 实现数据同步，确保写入。两个 NameNode 为了实现数据同步，需要通过一组 JournalNode 的独立进程进行通信。当活跃的 NameNode 中的 edits 日志有任何修改时，修改记录会写入大多数的 JournalNode；备用的 NameNode 监视并读取 JournalNode 中的变更数据，确保与活跃的 NameNode 数据同步。

由于 edits 日志的变更必须写入大多数（一半以上）JournalNode，所以至少应存在 3 个 JournalNode 进程，确保系统在单个主机出现故障时能够正常运行。一般设置 JournalNode 的数量为奇数。

HA 模式下的 Hadoop 集群一次只能有一个 NameNode 处于活跃状态，否则可能发生数据丢失或集群不能正常工作。事实上，为了防止脑裂的出现，JournalNode 只允许一个 NameNode 处于活跃状态，并对其进行相关的写操作。在故障转移期间，将要变为活跃状态的 NameNode 会全面接管写入 JournalNode 的操作，这有效地阻止了其他 NameNode 继续处于活跃状态。

（2）DFSZKFailoverController 是一个 ZooKeeper 客户端进程，主要负责监视 NameNode 的运行状态，管理 ZooKeeper 会话，进行基于 ZooKeeper 的选举。运行 NameNode 的每台服务器同时运行 DFSZKFailoverController 进程。

5. 验证自动故障转移

（1）查看 NameNode nn1 和 nn2 的状态。

查看 nn1 和 nn2 的状态有两种方法。

① 在虚拟机 qf01 上查看 nn1 和 nn2 的状态，首次查看状态用时十几秒，之后查看状态只需几秒。

```
[root@qf01 ~]# hdfs haadmin -getServiceState nn1
active
[root@qf01 ~]# hdfs haadmin -getServiceState nn2
```

```
standby
```

当前 nn1 处于活跃状态，nn2 处于备用状态。如果出现 nn1 处于备用状态，nn2 处于活跃状态的情况，也是正确的，因为 HA 模式下，HDFS 在启动时会随机指定一个 NameNode 处于活跃状态。

② 访问 NameNode Web 界面来查看 nn1 和 nn2 的状态。

在 Windows 的浏览器中，输入 http://192.168.142.131:50070，看到 nn1 处于活跃状态，如图 6.2 所示。

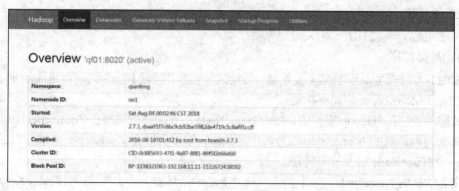

图 6.2　查看 nn1 的状态

在 Windows 的浏览器中，输入 http://192.168.142.132:50070，看到 nn2 处于备用状态，如图 6.3 所示。

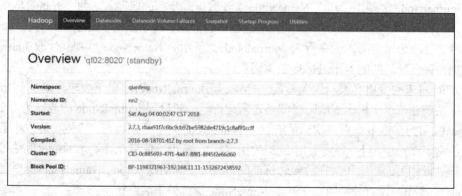

图 6.3　查看 nn2 的状态

（2）验证。

通过一些操作可以使活跃的 nn1 出现故障，查看是否可以成功将 nn2 切换到活跃状态。这些操作有：使用 kill 命令强制关闭 NameNode 进程来模拟虚拟机崩溃；通过关闭虚拟机或断开网络连接模拟不同类型的停机故障等。

本书以使用 kill 命令强制关闭 NameNode 进程为例，进行模拟操作，具体步骤如下。

① 在虚拟机 qf01 上，查看 NameNode 的进程 ID。

```
[root@qf01 ~]# jps
3447 NameNode
2408 QuorumPeerMain
3529 DataNode
17625 Jps
```

```
3660 JournalNode
3806 DFSZKFailoverController
```

② 使用 kill 命令强制关闭 NameNode 进程,使 nn1 出现故障。

```
[root@qf01 ~]# kill -9 3447
```

③ 在虚拟机 qf01 上查看 nn2 的状态。

```
[root@qf01 ~]# hdfs haadmin -getServiceState nn2
active
```

在几秒内 nn2 由原来的备用状态自动切换到活跃状态,表明 HDFS 实现了自动故障转移。至此,HDFS 的 HA 模式搭建完毕。

如果 nn2 未切换到活跃状态,则表明 HDFS 的 HA 模式搭建不成功,相关配置可能存在错误。

6.3.3 YARN 的 HA 模式

YARN 的 HA 模式架构如图 6.4 所示,具体搭建步骤如下。

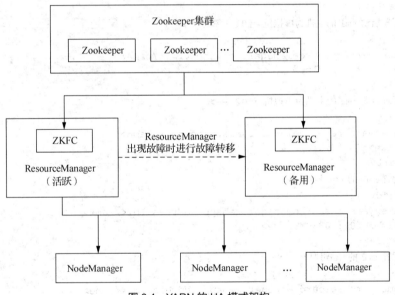

图 6.4 YARN 的 HA 模式架构

YARN 的 HA 模式下,有多个 ResourceManager,一个 ResourceManager 处于活跃(Active)状态,一个或多个 ResourceManager 处于备用(Standby)状态。活跃的 ResourceManager 负责将其状态写进 ZooKeeper,当该 ResourceManager 出现故障时,另一个备用的 ResourceManager 切换到活跃状态,切换方式有两种:管理员手动切换和配置自动故障转移。本书以配置自动故障转移为例,进行 YARN 的 HA 模式搭建。

1. YARN 的 HA 模式搭建

(1)确保 Hadoop 集群和 ZooKeeper 集群处于关闭状态。

(2)修改虚拟机 qf01 的/usr/local/hadoop-2.7.3/etc/hadoop/yarn-site.xml 文件,将 yarn-site.xml 文件中的内容替换如下。

```
<configuration>
```

```xml
    <!-- 配置 NodeManager 启动时加载 Shuffle 服务 -->
    <property>
        <name>yarn.nodemanager.aux-services</name>
        <value>mapreduce_shuffle</value>
    </property>
    <!-- 启动 yarn ResourceManager 的 HA 模式 -->
    <property>
        <name>yarn.resourcemanager.ha.enabled</name>
        <value>true</value>
    </property>
    <!-- 配置 yarn ResourceManager 的集群 ID -->
    <property>
        <name>yarn.resourcemanager.cluster-id</name>
        <value>qianfeng</value>
    </property>
    <!-- 指定 yarn ResourceManager 实现 HA 的节点名称 -->
    <property>
        <name>yarn.resourcemanager.ha.rm-ids</name>
        <value>rm1,rm2</value>
    </property>
    <!-- 配置启动 rm1 的主机为虚拟机 qf01 -->
    <property>
        <name>yarn.resourcemanager.hostname.rm1</name>
        <value>qf01</value>
    </property>
    <!-- 配置启动 rm2 的主机为虚拟机 qf02 -->
    <property>
        <name>yarn.resourcemanager.hostname.rm2</name>
        <value>qf02</value>
    </property>
    <!-- 配置 rm1 的 Web 地址 -->
    <property>
        <name>yarn.resourcemanager.webapp.address.rm1</name>
        <value>qf01:8088</value>
    </property>
    <!-- 配置 rm2 的 Web 地址 -->
    <property>
        <name>yarn.resourcemanager.webapp.address.rm2</name>
        <value>qf02:8088</value>
    </property>
    <!-- 配置 ZooKeeper 集群地址 -->
    <property>
        <name>yarn.resourcemanager.zk-address</name>
        <value>qf01:2181,qf02:2181,qf03:2181</value>
    </property>
</configuration>
```

（3）将虚拟机 qf01 上的 yarn-site.xml 文件分发到虚拟机 qf02、qf03。

```
[root@qf01 ~]# scp /usr/local/hadoop-2.7.3/etc/hadoop/yarn-site.xml qf02:/usr/local/hadoop-2.7.3/etc/hadoop/
[root@qf01 ~]# scp /usr/local/hadoop-2.7.3/etc/hadoop/yarn-site.xml qf03:/usr/local/hadoop-2.7.3/etc/hadoop/
```

（4）在虚拟机 qf01 的 /usr/local/hadoop-2.7.3/etc/hadoop/ 目录下新建文件 rm_hosts。

```
[root@qf01 ~]# vi /usr/local/hadoop-2.7.3/etc/hadoop/rm_hosts
```

在 rm_hosts 文件中输入如下内容。

```
qf01
qf02
```

（5）将虚拟机 qf01 的 /usr/local/hadoop-2.7.3/sbin/start-yarn.sh 文件的倒数第 5 行替换为如下内容。

```
"$bin"/yarn-daemons.sh --config $YARN_CONF_DIR --hosts rm_hosts start resourcemanager
```

（6）将虚拟机 qf01 的 /usr/local/hadoop-2.7.3/sbin/stop-yarn.sh 文件的倒数第 5 行替换为如下内容。

```
"$bin"/yarn-daemons.sh --config $YARN_CONF_DIR --hosts rm_hosts stop resourcemanager
```

（7）将虚拟机 qf01 的 rm_hosts、start-yarn.sh、stop-yarn.sh 文件分发到虚拟机 qf02、qf03。

```
[root@qf01 ~]# scp /usr/local/hadoop-2.7.3/etc/hadoop/rm_hosts qf02:/usr/local/hadoop-2.7.3/etc/hadoop/
[root@qf01 ~]# scp /usr/local/hadoop-2.7.3/etc/hadoop/rm_hosts qf03:/usr/local/hadoop-2.7.3/etc/hadoop/
[root@qf01 ~]# scp -r /usr/local/hadoop-2.7.3/sbin/ qf02:/usr/local/hadoop-2.7.3/sbin/
[root@qf01 ~]# scp -r /usr/local/hadoop-2.7.3/sbin/ qf03:/usr/local/hadoop-2.7.3/sbin/
```

2. 启动 YARN 的 HA 模式

先启动 ZooKeeper 集群，后启动所有节点的 YARN 进程。

```
[root@qf01 ~]# xzk.sh start
[root@qf01 ~]# start-yarn.sh
```

3. 验证 YARN 相关进程是否成功启动

在虚拟机 qf01 上查看虚拟机 qf01、qf02、qf03 的 YARN 相关进程。

```
[root@qf01 sbin]# xcmd.sh jps
============ qf01 jps ============
13588 ResourceManager
6037 QuorumPeerMain
13770 Jps
13707 NodeManager
============ qf02 jps ============
5904 QuorumPeerMain
11441 NodeManager
11359 ResourceManager
11551 Jps
============ qf03 jps ============
5717 QuorumPeerMain
10662 Jps
10604 NodeManager
```

出现以上进程，表明 YARN 相关进程成功启动。

4. 验证自动故障转移

（1）查看 rm1 和 rm2 的状态。

查看 rm1 和 rm2 的状态有两种方法。

① 在虚拟机 qf01 上查看 rm1 和 rm2 的状态，首次查看状态用时十几秒，之后查看状态只需几秒。

```
[root@qf01 ~]# yarn rmadmin -getServiceState rm1
standby
[root@qf01 ~]# yarn rmadmin -getServiceState rm2
active
```

当前 rm1 处于备用状态，rm2 处于活跃状态。与 HDFS 的 HA 模式类似，可能出现 rm1 处于活跃状态，rm2 处于备用状态的情况。

② 访问 ResourceManager Web 界面来查看 rm1 和 rm2 的状态。

在 Windows 的浏览器中，输入 http://192.168.142.131:8088，看到 rm1 处于备用状态，如图 6.5 所示。

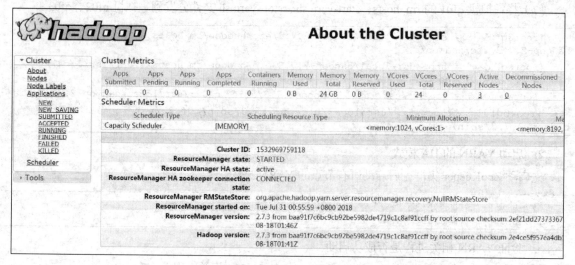

图 6.5　查看 rm1 的状态

在 Windows 的浏览器中，输入 http://192.168.142.132:8088，看到 rm2 处于活跃状态，如图 6.6 所示。

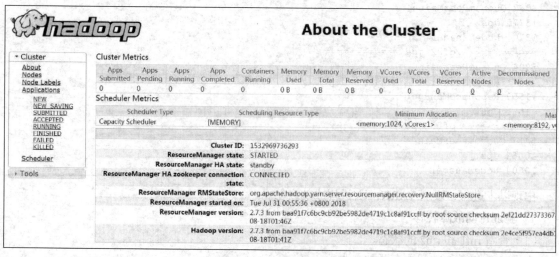

图 6.6　查看 rm2 的状态

（2）验证。

使用 kill 命令强制关闭 ResourceManager 进程，模拟自动故障转移，具体步骤如下。

① 在虚拟机 qf02 上，查看 ResourceManager 的进程 ID。

```
[root@qf02 ~]# jps
5904 QuorumPeerMain
11441 NodeManager
14174 Jps
11359 ResourceManager
```

② 使用 kill 命令强制关闭 ResourceManager 进程，使 rm2 出现故障。

```
[root@qf02 ~]# kill -9 11359
```

③ 查看 rm1 的状态。

```
[root@qf01 ~]# yarn rmadmin -getServiceState rm1
active
```

在几秒内 rm1 由原来的备用状态自动切换到活跃状态，表明 YARN 实现了自动故障转移。至此，YARN 的 HA 模式搭建完毕。

6.3.4 启动和关闭 Hadoop 的 HA 模式

（1）启动 Hadoop 的 HA 模式。在虚拟机 qf01 上，先启动 ZooKeeper 集群，后启动 Hadoop 集群。

```
[root@qf01 ~]# xzk.sh start
[root@qf01 ~]# start-dfs.sh
[root@qf01 ~]# start-yarn.sh
```

（2）查看相关进程。

```
[root@qf01 sbin]# xcmd.sh jps
============ qf01 jps ============
17316 NameNode
6037 QuorumPeerMain
17670 DFSZKFailoverController
18134 Jps
17767 ResourceManager
17863 NodeManager
17532 JournalNode
17405 DataNode
============ qf02 jps ============
5904 QuorumPeerMain
15985 ResourceManager
15250 NameNode
15620 JournalNode
16116 NodeManager
16600 Jps
15372 DataNode
15823 DFSZKFailoverController
============ qf03 jps ============
13890 DataNode
14594 Jps
```

```
 5717 QuorumPeerMain
14157 NodeManager
14014 JournalNode
```

（3）关闭 Hadoop 的 HA 模式。在虚拟机 qf01 上，先关闭 Hadoop 集群，后关闭 ZooKeeper 集群。

```
[root@qf01 ~]# stop-dfs.sh
[root@qf01 ~]# stop-yarn.sh
[root@qf01 ~]# xzk.sh stop
```

6.4 本章小结

本章主要讲解 Hadoop 2.0 的新特性。首先，讲解了 Hadoop 1.0 中 HDFS 和 MapReduce 存在的问题，以及 Hadoop 2.0 如何解决这些问题，让大家对 Hadoop 2.0 的新特性有基本的认识；其次，对 YARN 资源管理框架进行了讲解，让大家明白 YARN 是如何运行的；最后，讲解了 Hadoop 的高可用模式搭建。通过学习本章内容，应对 Hadoop 2.0 的新特性有进一步理解。

6.5 习题

1. 填空题

（1）Hadoop 的 HA 模式出现的原因是 Hadoop 集群存在_____问题。
（2）YARN 的主要组件包括_____、NodeManager、ApplicationMaster、Container。
（3）YARN 是基于 MapReduce 的_____与_____的框架。

2. 选择题

（1）（　　）不是 YARN 的优势。
  A. 更快的 MapReduce 计算　　　　B. 对多框架的支持
  C. 更高的容错性　　　　　　　　D. 框架升级更容易
（2）启动 Hadoop 的 HA 模式步骤正确的是（　　）。
  A. 先启动 ZooKeeper，后启动 Hadoop　　B. 先启动 Hadoop，后启动 ZooKeeper
  C. 只启动 ZooKeeper　　　　　　　　　　D. 只启动 Hadoop
（3）下面不属于 Hadoop 的 HA 模式启动后 qf03 机器进程的是（　　）。
  A. ZooKeeper　　　　　　　　　　B. NodeManager
  C. ResourceManager　　　　　　　D. DataNode

3. 思考题

（1）简述 HDFS 2.0 解决 HDFS 1.0 中单点故障和内存受限问题的方法。
（2）简述 YARN 的工作流程。

第7章 Hive

本章学习目标
- 熟悉 Hive 的安装
- 掌握 Hive 架构及其原理
- 掌握 Hive 的数据库和表的操作方法
- 熟悉 Hive 函数的使用
- 熟悉 Hive 的性能调优

Hive 是建立在 Hadoop 上的数据仓库工具，可以借助提取、转化、加载技术（Extract-Transform-Load，ETL）存储、查询和分析存储在 Hadoop 中的大规模数据。Hive 的出现使得开发人员使用相对简单的类 SQL（Struture Query Language，结构查询语言）语句，就可以操作 Hadoop 处理海量数据，大大降低了开发人员的学习成本。

7.1 数据仓库简介

7.1.1 数据仓库概述

数据仓库是一个面向主题的、集成的、随时间变化但信息本身相对稳定的数据集合，用于支持管理决策过程。总体来说，数据仓库可以整合多个数据源的历史数据，进行细粒度的、多维的分析，帮助高层管理者或者业务分析人员做出商业战略决策或商业报表。

7.1.2 数据仓库的使用

一个公司的不同项目可能用到不同的数据源，有的项目数据存在 MySQL 里面，有的项目数据存在 MongoDB 里面，甚至还有些要做第三方数据。

如果想把这些数据整合起来，进行数据分析，数据仓库（Data Warehouse，DW）就派上用场了。它可以对多种业务数据进行筛选和整合，用于数据分析、数据挖掘、数据报表，如图 7.1 所示。

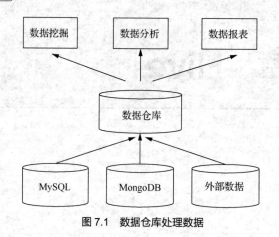

图 7.1　数据仓库处理数据

7.1.3　数据仓库的特点

（1）主题性：数据仓库针对某个主题来进行组织，比如"滴滴出行"的司机行为分析就是一个主题，所以它可以对多种不同的数据源进行整合；而传统的数据库主要针对某个项目，数据相对分散和孤立。

（2）集成性：数据仓库需要将多个数据源的数据存到一起，但是这些数据以前的存储方式不同，所以要经过抽取、清洗、转换的过程。

（3）稳定性：数据仓库保存的数据是一系列历史快照，不允许修改，只能分析。

（4）时变性：数据仓库会定期接收到新的数据，反映最新的数据变化。

7.1.4　主流的数据仓库

国内最常用的是一款基于 Hadoop 的开源数据仓库，名为 Hive，它可以对存储在 HDFS 的文件数据进行查询、分析。

Hive 对外可以提供 HQL（Hive Query Language，Hive 查询语言），这是类似于 SQL 的一种查询语言。在查询时 HQL 语句可转换为 MapReduce 任务，在 Hadoop 层执行。

Hive 的最大优势在于免费存储海数据，其他知名的商业数据仓库还有 Oracle、DB2，以及业界领先的 Teradata。

Teradata 支持大规模并行处理，可以高速处理海量数据，性能远远高于 Hive。使用 Teradata 的企业只需要专注于业务，能够节省管理方面的精力，实现投资回报率最大化。

7.2　认识 Hive

7.2.1　Hive 简介

1. Hive 诞生的背景

开发人员在使用 MapReduce 的过程中面临着以下两个问题。

（1）Hadoop 的 MapReduce 专业性较强，学习成本相对较高。

（2）通过 MapReduce 实现复杂查询等操作时，开发难度相对较大。

为解决以上问题，使用类 SQL 语法的 Hive 应运而生。Hive 诞生于 Facebook，大量懂得 SQL 的开发人员快速学会通过 Hive 操作 Hadoop 集群处理海量数据,满足了 Facebook 管理海量社交数据和进行机器学习的需求。Hive 是学习 Hadoop 相关技术的一个突破口。

2. Hive 的概念

Hive 是基于 Hadoop 的数据仓库工具。Hive 提供了操作简单的类 SQL 查询语言 HQL,HQL 语句可以转换为 MapReduce 任务来执行,方便用户查询、分析数据;Hive 还提供了命令行工具和 JDBC (Java DataBase Connectivity,Java 数据库连接)驱动程序以便用户连接到 Hive。

数据仓库是为企业做出科学的决策报告而创建的,可以对多样的业务数据进行筛选与整合的大型数据存储集合。用户一般会利用数据仓库技术执行大量的查询操作,执行修改和删除操作较少,因此数据仓库通常只需要定期加载和刷新。

3. Hive 与 Hadoop 的关系

Hive 的构建基于 Hadoop,Hive 与 Hadoop 的关系具体体现在以下几个方面。

(1) Hive 处理的真实数据存储在 Hadoop 文件系统中,通常是 HDFS。

(2) Hive 的大部分查询操作通过 MapReduce 作业完成(只有小部分查询操作没有使用 MapReduce 作业,如 select * from table)。

(3) Hadoop 和 Hive 都使用 UTF-8 编码。

4. Hive 的优缺点

(1) Hive 的优点如下。

- 提供了易于操作的类 SQL 查询语言 HQL,学习成本相对较低。
- 利用 MapReduce 作为计算引擎,HDFS 作为存储系统,可以操作大数据集。
- 支持用户自定义函数,用户可以根据自己的需求编写函数。
- 具有良好的容错性,一个节点出现问题仍可完成相关操作。

(2) Hive 的缺点如下。

- HQL 语言处理复杂业务的能力有限。
- Hive 的运行效率较低。
- Hive 不提供实时查询功能。

7.2.2 Hive 架构

以 Hive2.1.1 版本为例,Hive 的架构如图 7.2 所示。

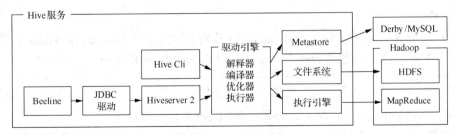

图 7.2 Hive 架构图

(1) Hive Cli。Hive Cli 是 Hive 1.0 的命令行客户端(Shell 环境)。

(2) Beeline。Beeline 是 Hive 新的命令行客户端,类似于 Hive Cli,功能更强大,支持嵌入模式和远程模式(使用 JDBC 驱动)连接 Hiveserver2 服务来操作 Hive。在 Hive 2.1.1 版本中推荐使用 Beeline 客户端。

(3) Hiveserver2。Hiveserver2 是一种使 Hive 客户端能够与 Hive 进行通信的 Hive 服务。

(4) 驱动引擎。驱动引擎主要包括解释器、编译器、优化器、执行器。通过驱动引擎能够完成 HQL 查询语句的语法解释、编译、优化、查询计划生成。

- 解释器用来将 Hive 语句转换为 MapReduce 程序。
- 编译器用来将转换后的 MapReduce 程序打成 jar 包。

(5) Metastore。Hive 客户端连接 Metastore 服务，Metastore 服务连接关系型数据库来存储元数据。Hive 中的元数据包括表的名称、表的列和分区及其属性、表的属性（是否为外部表等）、表的数据所在目录等。Metastore 连接的关系型数据库一般是 Derby 或 MySQL。Derby 是 Hive 默认使用的存储元数据的数据库，只能支持一个 Hive 会话。在实际开发环境中，需要使用多个 Hive 会话进行 Hive 的相关操作，而 MySQL 可以支持多个 Hive 会话，所以在企业中使用 MySQL 存储 Hive 元数据的情况比较常见。本书以 MySQL 为例进行讲述。

7.2.3 Hive 和关系型数据库比较

Hive 采用的 HQL 和关系型数据库（Relational Database Management System，RDBMS）的 SQL 类似，因此容易将 Hive 理解为数据库。实际上，Hive 和 RDBMS 区别较大，较突出的区别是 Hive 不适合用于联机（Online）事务处理，不提供实时查询功能，而 RDBMS 可以用在 Online 的应用中。Hive 和 RDBMS 对比，如表 7.1 所示。

表 7.1　　　　　　　　　　　　　　Hive 和 RDBMS 对比

对比项	Hive	RDBMS
查询语言	HQL	SQL
数据存储	HDFS	块设备或本地文件系统
执行	MapReduce	Executor
执行延迟	较高	较低
处理数据规模	较大	较小
可扩展性	较高	较低

7.3　Hive 安装

本书使用 2.1.1 版本的 Hive，安装包见附录。Hive 的具体安装步骤如下。

1. 安装 Hive

（1）将 Hive 安装包 apache-hive-2.1.1-bin.tar.gz 放到虚拟机 qf01 的 /root/Downloads/ 目录下，切换到 root 用户，解压 Hive 安装包到 /mysoft 目录（该目录在安装 ZooKeeper 时已创建）下。

```
[root@qf01 ~]# tar -zxvf /root/Downloads/apache-hive-2.1.1-bin.tar.gz -C /mysoft/
```

（2）切换到 /mysoft 目录下，将 apache-hive-2.1.1-bin 重命名为 hive。

```
[root@qf01 ~]# cd /mysoft/
[root@qf01 mysoft]# mv apache-hive-2.1.1-bin hive
```

（3）打开 /etc/profile 文件，配置 Hive 环境变量。

```
[root@qf01 mysoft]# vi /etc/profile
```

在文件末尾添加如下三行内容。

```
# Hive environment variables
export HIVE_HOME=/mysoft/hive
export PATH=$PATH:$HIVE_HOME/bin
```

(4) 使环境变量生效。

```
[root@qf01 mysoft]# source /etc/profile
```

2. 安装 MySQL

本书默认使用 5.6 版本的 MySQL。

(1) 将与 MySQL 相关的两个文件 mysql-community-release-el7-5.noarch.rpm、mysql-connector-java-5.1.41.jar 和 mysql 目录（见附录）放到虚拟机 qf01 的/root/Downloads/目录下。

(2) 切换到/root/Downloads/目录下，安装 mysql-community-release-el7-5.noarch.rpm 文件。

```
[root@qf01 mysoft]# cd /root/Downloads/
[root@qf01 Downloads]# rpm -ivh mysql-community-release-el7-5.noarch.rpm
```

(3) 切换到 mysql 目录，安装 MySQL。

```
[root@qf01 Downloads]# cd mysql
[root@qf01 mysql]# yum localinstall *
```

(4) 启动 MySQL 进程，将其设置为开机启动。

```
[root@qf01 mysql]# chkconfig --add mysql
[root@qf01 mysql]# chkconfig mysql on
[root@qf01 mysql]# service mysql start
```

(5) 查看 MySQL 进程是否启动成功。

```
[root@qf01 mysql]# service mysql status
```

(6) 进入 MySQL。

```
[root@qf01 mysql]# mysql
```

(7) 设置 MySQL 的 root 用户的密码为 root。

```
mysql> update mysql.user set password=password('root');
```

(8) 更新权限。

```
mysql> flush privileges;
```

(9) 退出 MySQL。

```
mysql> exit
```

(10) 配置密码后，进入 MySQL 的命令如下。

```
[root@qf01 mysql]# mysql -uroot -p
Enter password:
...
mysql>
```

提示：出现 Enter password 后，输入 root。

至此，MySQL 安装完成。

3. 配置 Hive

通过配置 Hive，将 Hive 的存储 Metastore 数据的数据库由 Derby 更换为 MySQL。

（1）切换到/mysoft/hive/conf 目录，修改如下两个配置文件的名称。

```
[root@qf01 mysql]# cd /mysoft/hive/conf
[root@qf01 conf]# mv hive-env.sh.template hive-env.sh
[root@qf01 conf]# mv hive-default.xml.template hive-site.xml
```

（2）修改 hive-env.sh 文件。将# HADOOP_HOME=${bin}/../../hadoop 一行替换为如下内容。

```
HADOOP_HOME=/usr/local/hadoop-2.7.3
```

（3）修改 hive-site.xml 文件。

① 修改 hive-site.xml 文件中的 MySQL 数据库信息：驱动、连接、账号、密码。查找以下 4 行内容。

```
<value>org.apache.derby.jdbc.EmbeddedDriver</value>
<value>jdbc:derby:;databaseName=metastore_db;create=true</value>
<value>APP</value>
<value>mine</value>
```

依次替换为以下 4 行内容。

```
<value>com.mysql.jdbc.Driver</value>
<value>jdbc:mysql://qf01:3306/hive?createDatabaseIfNotExist=true</value>
<value>root</value>
<value>root</value>
```

② 使用 vi 文本编辑器的替换字符串功能，将 hive-site.xml 文件中的${system:java.io.tmpdir}替换为/root/hivetemp，${system:user.name}替换为 root。

```
:%s#${system:java.io.tmpdir}#/root/hivetemp#g
:%s#${system:user.name}#root#g
```

（4）将 MySQL 的驱动文件 mysql-connector-java-5.1.41.jar 复制到/mysoft/hive/lib 目录下。

```
[root@qf01 conf]# cp /root/Downloads/mysql-connector-java-5.1.41.jar /mysoft/hive/lib
```

（5）初始化 Metastore。

```
[root@qf01 conf]# schematool -initSchema -dbType mysql
```

4. 启动 Hive 的 Cli 客户端

（1）在启动 Hive 的 Cli 客户端之前，需要先启动 Hadoop 的 HA 模式。

```
[root@qf01 conf]# xzk.sh start
[root@qf01 conf]# start-dfs.sh
[root@qf01 conf]# start-yarn.sh
[root@qf01 conf]# hive
...
hive>
```

（2）进行简单的测试，验证 Hive 是否安装成功。

① 查看数据库。

```
hive> show databases;
OK
```

```
default
Time taken: 1.388 seconds, Fetched: 1 row(s)
```

② 创建数据库，并查看数据库是否成功创建。

```
hive> create database if not exists qfdb01;
hive> show databases;
OK
qfdb01
default
Time taken: 0.012 seconds, Fetched: 2 row(s)
```

如果两个测试的返回结果如上所示，表明 Hive 安装成功。

（3）退出 Hive 的 Cli 客户端。

```
hive> exit;
```

5. 启动 Hive 的 Beeline 客户端

（1）关闭 Hadoop 的 HDFS 和 YARN 进程，ZooKeeper 的相关进程不用关闭。

```
[root@qf01 ~]# stop-dfs.sh
[root@qf01 ~]# stop-yarn.sh
```

（2）在虚拟机 qf01 上修改 /usr/local/hadoop-2.7.3/etc/hadoop 目录下的 core-site.xml 文件。

```
[root@qf01 ~]# vi /usr/local/hadoop-2.7.3/etc/hadoop/core-site.xml
```

（3）在 core-site.xml 中的 </configuration> 一行之前，添加如下几行内容。

```xml
<property>
    <name>hadoop.proxyuser.root.hosts</name>
    <value>*</value>
</property>
<property>
    <name>hadoop.proxyuser.root.groups</name>
    <value>*</value>
</property>
```

（4）将 core-site.xml 文件分发到虚拟机 qf02、qf03。

```
[root@qf01 ~]# scp /usr/local/hadoop-2.7.3/etc/hadoop/core-site.xml qf02:/usr/local/hadoop-2.7.3/etc/hadoop/
[root@qf01 ~]# scp /usr/local/hadoop-2.7.3/etc/hadoop/core-site.xml qf03:/usr/local/hadoop-2.7.3/etc/hadoop/
```

（5）启动 Hadoop 的 HDFS 和 YARN 进程。

```
[root@qf01 ~]# start-dfs.sh
[root@qf01 ~]# start-yarn.sh
```

（6）在虚拟机 qf01 上启动 Hive 的 Hiveserver2 服务。

```
[root@qf01 ~]# hiveserver2
...
SLF4J: Actual binding is of type [org.apache.logging.slf4j.Log4jLoggerFactory]
```

启动 Hive 的 Hiveserver2 服务后，命令行处于塞住状态，等待客户端的接入。

（7）在虚拟机 qf01 上，重新打开一个终端窗口，启动 Hive 的 Beeline 客户端。

```
[root@qf01 ~]# beeline
...
beeline>
```

（8）Beeline 使用 JDBC 驱动连接 Hiveserver2 服务。

Beeline 使用 JDBC 驱动连接 Hiveserver2 服务的方法有以下两种。

① 分步骤连接。

```
beeline> !connect jdbc:hive2://localhost:10000
```

提示：出现 Enter username for jdbc:hive2://localhost:10000:时，按回车键；出现 Enter password for jdbc:hive2://localhost:10000:时，按回车键。

出现以下内容，表明成功连接 Hiveserver2 服务。

```
0: jdbc:hive2://localhost:10000>
```

注意：Hiveserver2 默认使用的端口号是 10000。

② 在 CentOS 的命令行输入以下命令直接连接。

```
[root@qf01 ~]# beeline -u jdbc:hive2://localhost:10000 -n root
...
0: jdbc:hive2://localhost:10000>
```

本书为方便表述，以下使用 jdbc:hive2://>代替 0: jdbc:hive2://localhost:10000>。

7.4 Hive 数据类型

Hive 的数据类型分为基本数据类型和复杂数据类型。

7.4.1 Hive 基本数据类型

Hive 常见基本数据类型，如表 7.2 所示。

表 7.2　　　　　　　　　　　　　　Hive 常见基本数据类型

基本数据类型		描述	示例
数字类型	TINYINT	1 字节有符号整数，从 128 至 127	3Y
	SMALLINT	2 字节有符号整数，从-32 768 至 32 767	3S
	INT / INTEGER	4 字节有符号整数，从-2 147 483 648 至 2 147 483 647	3
	BIGINT	8 字节有符号整数，从-9 223 372 036 854 775 808 至 9 223 372 036 854 775 807	3L
	FLOAT	4 字节单精度浮点数	3.0
	DOUBLE	8 字节双精度浮点数	3.0
日期/时间类型	TIMESTAMP	从 Hive 0.8.0 开始提供	1528236366000,'2018-06-06 06:06:06'
	DATE	从 Hive 0.12.0 开始提供	'2018-06-06'
字符串类型	STRING	常规字符串	'qf', "qf"
	VARCHAR	字符数介于 1 和 65 355 之间，从 Hive 0.12.0 开始提供	'qf', "qf"
	CHAR	最大字符数为 255，从 Hive 0.13.0 开始提供	'qf', "qf"
布尔类型	BOOLEAN	true/false	true
二进制类型	BINARY	用于存储变长的二进制数据，从 Hive 0.8.0 开始提供	

Hive 是使用 Java 语言开发的，Hive 里的基本数据类型和 Java 的基本数据类型大致上是一一对应的。Hive 中有符号的整数类型 TINYINT、SMALLINT、INT 和 BIGINT 分别对应 Java 的 byte、short、int 和 long 类型。Hive 的浮点数据类型 FLOAT 和 DOUBLE 分别对应 Java 的 float 和 double 类型。而 Hive 的 BOOLEAN 类型对应 Java 的 boolean 类型。

Hive 的基本数据类型可以进行隐式转换，TINYINT 类型能够自动转为 INT 类型，但是 INT 不能自动转为 TINYINT 类型；TINYINT、SMALLINT、INT、INTEGER 类型都可以转为 FLOAT 类型；TINYINT、SMALLINT、INT、INTEGER、BIGINT、FLOAT 和 STRING 类型都可以转为 DOUBLE 类型；BOOLEAN 类型不可以转换为其他基本数据类型。

7.4.2 Hive 复杂数据类型

复杂数据类型可以使用基本数据类型和其他复合类型来进行构建，Hive 复杂数据类型包括 ARRAY、MAP、STRUCT、UNIONTYPE。其中，ARRAY、MAP 与 Java 中的 Array 和 Map 相似，STRUCT 与 C 语言中的 Struct 相似。复杂数据类型允许任意层次的嵌套。在大数据系统中，使用复杂数据类型数据的好处在于能够减少寻址次数，提高查询速度，提高数据的吞吐量。

下面对复杂数据类型分别进行讲解。

1. ARRAY

ARRAY 中的元素必须是相同类型。可以使用下标索引访问 ARRAY 的元素，例如，ARRAY 类型的数据 A 为{'x', 'y', 'z'}，其中，A[0]为'x'。

2. MAP

MAP 类型的数据以键值对的形式存在，使用[元素名称]表示法访问元素，即通过键访问值。键必须是基本数据类型，值可以是任意数据类型。例如，MAP 类型的数据 M 为{name:'tom',age: '18'}，M[name]访问的是'tom'。

3. STRUCT

STRUCT 是用户自定义的结构，需要自定义字段。可以使用.（下圆点）表示法访问 STRUCT 类型的数据中的元素。

4. UNIONTYPE

UNIONTYPE 是从 Hive 0.7.0 开始支持的。UNIONTYPE 可以在定义的多个数据类型中任意指定一个。

接下来通过一个示例来理解 Hive 复杂数据类型的使用方法，涉及的 Hive 数据库和表的相关操作在后面章节会进行详细讲解。

（1）准备数据。

在虚拟机 qf01 的/root 目录下新建文件 complexdata.txt，具体数据如下。

```
Sophie|London,Britain|Female,30|Java:80|Developmen:01
Coco|Paris,France|Female,38|Python:85|Product:02
Tom|Washington,America|Male,27|Kotlin:80|HR:03
Jack|Beijing,China|Male,57|Scala:89|Test:04
```

（2）创建表。

Hive 默认自带一个名为 default 的数据库，如果新建表时没有指定存放的数据库，Hive 默认将表创建到 default 数据库中。

```
jdbc:hive2://> create table complex(
```

```
co1 string,
co2 array<string>,
co3 struct<sex:string,age:int>,
co4 map<string,int>,
co5 uniontype<string,int>
)
row format delimited
fields terminated by '|'
collection items terminated by ','
map keys terminated by ':'
lines terminated by '\n'
stored as textfile;
```

其中,co1、co2、co3、co4、co5 是各列的字段名称。

(3)将 complexdata.txt 文件加载到 Hive 的 default 数据库的 complex 表中。

```
jdbc:hive2://> load data local inpath '/root/complexdata.txt' into table complex;
```

(4)查询复杂数据类型的字段。

```
jdbc:hive2://> select co1,co2[0],co3.sex,co4['Java'],co5 from complex;
+---------+-------------+---------+-------+-------------------------------------+--+
|co1      | _c1         |sex      |_c3    |co5                                  |
+---------+-------------+---------+-------+-------------------------------------+--+
|Sophie|London|Female|80|{0:"Sophie|London,Britain|Female,30|Java:80|Developmen:01"}|
|Coco  |Paris |Female|NULL|{0:"Coco|Paris,France|Female,38|Python:85|Product:02"}|
|Tom   |Washington|Male|NULL|{0:"Tom|Washington,America|Male,27|Kotlin:80|HR:03"}|
|Jack  |Beijing|Male|NULL|{0:"Jack|Beijing,China|Male,57|Scala:89|Test:04"}|
+---------+-------------+---------+-------+-------------------------------------+--+
4 rows selected (3.214 seconds)
```

其中,查询 co4['Java']时,complexdata.txt 文件的后 3 行数据中没有对应的数据,返回 NULL。

7.5 Hive 数据库操作

Hive 的表都存储在数据库里面,因此需要先了解 Hive 的数据库操作。Hive 对数据库的操作与 MySQL 类似,常见操作如下。

(1)查看所有的数据库。

```
jdbc:hive2://> show databases;
```

(2)使用指定的数据库。

```
jdbc:hive2://> use default;
```

(3)查看数据库信息。

```
jdbc:hive2://> desc database default;
```

Hive 会为每一个数据库创建一个目录,这个数据库中的表将会以子目录的形式放在这个数据库目录下。在创建数据库时,如果没有指定存储位置,默认存放在 HDFS 的/user/hive/warehouse 目录下。default 数据库中的表直接存放在 HDFS 的/user/hive/warehouse 目录下。

(4)创建数据库。

```
jdbc:hive2://> create database qfdb02;
```

（5）切换到数据库 qfdb02。

```
jdbc:hive2://> use qfdb02;
```

（6）查看当前使用的数据库。

```
jdbc:hive2://> select current_database();
+--------+--+
|  _c0   |
+--------+--+
| qfdb02 |
+--------+--+
1 row selected (0.165 seconds)
```

（7）删除数据库 qfdb02。

```
jdbc:hive2://> drop database qfdb02;
```

删除含有数据的数据库时会报错。如果要强制删除数据库，需要添加 cascade 关键字，如下所示。

```
jdbc:hive2://> drop database if exists qfdb02 cascade;
```

提示：Beeline 使用 JDBC 连接 Hiveserver2，默认使用的数据库是 default。

7.6 Hive 表

Hive 的表在逻辑上由存储的真实数据和元数据组成。真实数据一般存放在 HDFS 中，元数据存放在关系型数据库中。

Hive 的表主要有内部表（又称托管表）和外部表，还有在表中进行分区得到的分区表、在表或分区中分桶得到的桶表。

7.6.1 内部表和外部表

简单地理解，创建时不使用 external 关键字的表就是内部表，创建时使用 external 关键字的表就是外部表。

1. 常用的建表规则

在学习内部表和外部表之前，先来了解常用的建表规则（[]中的内容为可选内容）。

```
CREATE [TEMPORARY] [EXTERNAL] TABLE [IF NOT EXISTS] [db_name.]table_name
  [(col_name data_type [COMMENT col_comment], ... [constraint_specification])]
  [COMMENT table_comment]
  [
   [ROW FORMAT row_format]
   [STORED AS file_format]
  ]
  [LOCATION hdfs_path]
```

（1）create table 用于创建一个指定名称的表。如果已经存在该名称的表，则抛出异常；用户可以用 if not exists 选项来忽略这个异常。

（2）external 关键字可以让用户创建一个外部表。

（3）comment 能够为表与字段添加描述。

（4）row format 用于指定表的行格式，常见使用方法如下。

```
ROW FORMAT DELIMITED
[FIELDS TERMINATED BY char]
[COLLECTION ITEMS TERMINATED BY char]
[MAP KEYS TERMINATED BY char]
[LINES TERMINATED BY char]
```

（5）stored as 用于指定表存储的文件格式。Hive 常见的文件格式有 TextFile、SequenceFile、RCFile、ORCFile、Parquet。Hive 默认的文件格式是 TextFile。

（6）location 用于指定外部表的存储位置，一般是 HDFS 的目录。

2. 内部表和外部表的创建

（1）创建内部表。

创建内部表，示例如下。

```
jdbc:hive2://> create table internal_table(
co1 string,
co2 array<string>,
co3 struct<sex:string,age:int>,
co4 map<string,int>,
co5 uniontype<string,int>
)
comment 'this is a internal_table'
row format delimited
fields terminated by '|'
collection items terminated by ','
map keys terminated by ':'
lines terminated by '\n'
stored as textfile;
```

其中，name 是列的字段名称，fields terminated by '|'规定字段与字段之间的分隔符为|（Hive 默认的字段分隔符为\001），collection items terminated by ','规定一个字段各个条目的分隔符为,。

提示：Hive 创建表时，如果未指定存放在哪个数据库，默认存放在 default 数据库中。

查看所有的表，查看内部表是否创建成功。

```
jdbc:hive2://> show tables;
+----------------+--+
| tab_name       |
+----------------+--+
| complex        |
| internal_table |
+----------------+--+
2 rows selected (0.06 seconds)
```

由以上结果可知，内部表 internal_table 已在 Hive 中。

查看表结构（字段和数据类型）。

```
jdbc:hive2://> desc internal_table;
+----------+---------------------------+---------+--+
| col_name | data_type                 | comment |
+----------+---------------------------+---------+--+
| co1      | string                    |         |
| co2      | array<string>             |         |
| co3      | struct<sex:string,age:int>|         |
```

```
| co4        | map<string,int>       |       |    |
| co5        | uniontype<string,int> |       |    |
+------------+-----------------------+-------+----+
5 rows selected (0.168 seconds)
```

查看表的属性。

```
jdbc:hive2://> desc formatted internal_table;
```

（2）创建外部表。

创建外部表，示例如下。

① 创建外部表的 HDFS 存储目录。

外部表的存储位置需要在创建外部表时指明。在虚拟机 qf01 上，新建 HDFS 级联目录 /user/qf/external_table。

```
[root@qf01 ~]# hdfs dfs -mkdir -p /user/qf/external_table
```

② 创建外部表。

```
jdbc:hive2://> create external table external_table(name string)
location '/user/qf/external_table';
```

在实际开发环境中，如果所有操作都由 Hive 完成，一般选择使用内部表。如果 Hive 和其他工具使用同一个数据集进行操作，一般选择使用外部表。

3. 加载数据到表中

内部表和外部表的区别主要体现在加载表和删除表上。

（1）加载数据到内部表。

加载数据到内部表有以下两种情况。

① 加载虚拟机 qf01 的/root 目录下的 complexdata.txt 文件到内部表。

```
jdbc:hive2://> load data local inpath '/root/complexdata.txt' into table complex;
```

② 加载 HDFS 的/user/root 目录下的 complexdata.txt 文件到内部表，需要先上传 complexdata.txt 文件到 HDFS 的/root 目录下（可以重新打开一个终端窗口进行该操作）。

```
[root@qf01 ~]# hdfs dfs -put complexdata.txt /user/root
jdbc:hive2://> load data inpath '/user/root/complexdata.txt' into table complex;
```

该操作是把 HDFS 的/user/root 目录下的 complexdata.txt 文件移动到 Hive 的 complex 表的仓库目录下，Hive 的 complex 表的仓库目录为 HDFS 的/user/hive/warehouse/complex 目录。

（2）加载数据到外部表。

```
[root@qf01 ~]# hdfs dfs -put complexdata.txt /user/root
jdbc:hive2://> load data inpath '/user/root/complexdata.txt' into table external_table;
```

该操作不会把 HDFS 的/user/root 目录下的 complexdata.txt 文件移动到 Hive 的 external_table 表的仓库目录下。

4. 删除表

删除表的一般操作如下。

```
jdbc:hive2://> drop table external_table;
jdbc:hive2://> drop table internal_table;
```

注意：删除内部表，会把关系型数据库中存储的元数据和 HDFS 中存储的真实数据一起删除。删除外部表，只删除关系型数据库中存储的元数据，HDFS 中存储的真实数据仍然存在。

5. 复制表

Hive 复制表时，常用的关键字是 like 和 as。

（1）like 关键字。

like 关键字只用来复制指定表的表结构，并不复制其数据。

使用 like 关键字创建表 copy_insert_data。

```
jdbc:hive2://> create table copy_insert_data like insert_data;
```

分别查看 copy_insert_data 和 insert_data 的表结构。

```
jdbc:hive2://> desc copy_insert_data;
+-----------+------------+----------+--+
| col_name  | data_type  | comment  |
+-----------+------------+----------+--+
| name      | string     |          |
+-----------+------------+----------+--+
1 row selected (0.099 seconds)


+-----------+------------+----------+--+
| col_name  | data_type  | comment  |
+-----------+------------+----------+--+
| name      | string     |          |
+-----------+------------+----------+--+
1 row selected (0.168 seconds)
```

由返回结果可知，copy_insert_data 复制了 insert_data 的表结构。

查看 copy_insert_data 的数据。

```
jdbc:hive2://> select * from copy_insert_data;
+------------------------+--+
| copy_insert_data.name  |
+------------------------+--+
+------------------------+--+
No rows selected (0.162 seconds)
```

由查询结果可知，copy_insert_data 并没有复制 insert_data 的数据。

（2）as 关键字。

as 关键字用来复制指定表的表结构和数据。

使用 as 关键字创建表 copy_as_insert_data。注意：使用 as 关键字创建表的语句和 like 略有差别。

```
jdbc:hive2://> create table copy_as_insert_data as select * from insert_data;
```

查看 copy_as_insert_data 的表结构。

```
jdbc:hive2://> desc copy_as_insert_data;
+-----------+------------+----------+--+
| col_name  | data_type  | comment  |
+-----------+------------+----------+--+
| name      | string     |          |
+-----------+------------+----------+--+
1 row selected (0.103 seconds)
```

查看 copy_as_insert_data 的数据。

```
jdbc:hive2://> select * from copy_as_insert_data;
+---------------------------+--+
| copy_as_insert_data.name  |
+---------------------------+--+
| Sophie                    |
+---------------------------+--+
1 row selected (0.243 seconds)
```

由以上查询结果可知，copy_as_insert_data 复制了指定表的表结构和数据。

在创建表时，as 关键字还可以用来复制指定表的指定列。

创建表。

```
jdbc:hive2://> create table t1 as select new_complex.co1, new_complex.co2 from new_complex;
```

查看 t1 的表结构。

```
jdbc:hive2://> desc t1;
+-----------+---------------+----------+--+
| col_name  |   data_type   | comment  |
+-----------+---------------+----------+--+
| co1       | string        |          |
| co2       | array<string> |          |
+-----------+---------------+----------+--+
2 rows selected (0.1 seconds)
```

查看 t1 的数据。

```
jdbc:hive2://> select * from t1;
+---------+-----------------------------+--+
| t2.co1  |          t2.co2             |
+---------+-----------------------------+--+
| Sophie  | ["London","Britain"]        |
| Coco    | ["Paris","France"]          |
| Tom     | ["Washington","America"]    |
| Jack    | ["Beijing","China"]         |
+---------+-----------------------------+--+
4 rows selected (0.18 seconds)
```

6. 删除表的数据，保留表结构

truncate 关键字用来删除表的数据，保留表结构。

删除表数据。

```
jdbc:hive2://> truncate table t1;
```

查看 t1 的数据。

```
jdbc:hive2://> select * from t1;
+---------+---------+--+
| t1.co1  | t1.co2  |
+---------+---------+--+
+---------+---------+--+
No rows selected (1.59 seconds)
```

返回结果表明表 t1 中的数据已经被删除。

查看 t1 的表结构。

```
jdbc:hive2://> desc t1;
+-----------+------------------+----------+--+
| col_name  |    data_type     | comment  |
+-----------+------------------+----------+--+
| co1       | string           |          |
| co2       | array<string>    |          |
+-----------+------------------+----------+--+
2 rows selected (0.154 seconds)
```

返回结果表明表 t1 的结构依然存在。

7. 修改表

（1）修改表名。

```
jdbc:hive2://> alter table complex rename to new_complex;
jdbc:hive2://> show tables;
+--------------------+--+
|      tab_name      |
+--------------------+--+
| external_table     |
| internal_table     |
| new_complex        |
+--------------------+--+
3 rows selected (0.03 seconds)
```

（2）修改表注释。

```
jdbc:hive2://> alter table internal_table set tblproperties('comment' = 'this is a new comment');
jdbc:hive2://> desc formatted internal_table;
```

（3）修改列。

① 增加列。

```
jdbc:hive2://> alter table internal_table add columns(co6 string, co7 int);
jdbc:hive2://> desc internal_table;
+-----------+------------------------------+------+--+
| col_name  |          data_type           |      |
+-----------+------------------------------+------+--+
| co1       | string                       |      |
| co2       | array<string>                |      |
| co3       | struct<sex:string,age:int>   |      |
| co4       | map<string,int>              |      |
| co5       | uniontype<string,int>        |      |
| co6       | string                       |      |
| co7       | int                          |      |
+-----------+------------------------------+------+--+
7 rows selected (0.059 seconds)
```

② 修改列名、列的类型、列的注释、列的位置，通用格式如下。

```
ALTER TABLE table_name CHANGE
[CLOUMN] col_old_name col_new_name column_type
[CONMMENT col_conmment]
[FIRST|AFTER column_name];
```

first 用来将列放在第 1 列，after column_name 用来将指定列放在 col_name 后面。

示例：将表 internal_table 的 co7 列名改为 co9，将列的类型 int 改为 string，添加列的注释，将列的位置移动到 co5 之后。

```
jdbc:hive2://> alter table internal_table change co7 co9 string comment 'the datatype of co8 is string' after co5;
jdbc:hive2://> desc internal_table;
+----------+-----------------------------+----------------------------+--+
| col_name |          data_type          |          comment           |
+----------+-----------------------------+----------------------------+--+
| co1      | string                      |                            |
| co2      | array<string>               |                            |
| co3      | struct<sex:string,age:int>  |                            |
| co4      | map<string,int>             |                            |
| co5      | uniontype<string,int>       |                            |
| co9      | string                      | the datatype of co9 is string |
| co6      | string                      |                            |
+----------+-----------------------------+----------------------------+--+
7 rows selected (0.077 seconds)
```

③ 删除列。以删除 co6 列为例，具体操作如下。

```
jdbc:hive2://> alter table internal_table replace columns(
co1 string,
co2 array<string>,
co3 struct<sex:string,age:int>,
co4 map<string,int>,
co5 uniontype<string,int>,
co9 string comment 'the datatype of co9 is string'
);
```

8. 向表中插入数据

创建表 insert_data，插入数据，并查询插入的数据。

```
jdbc:hive2://> create table insert_data(name string);
jdbc:hive2://> insert into insert_data values('Sophie');
jdbc:hive2://> select * from insert_data;
+-------------------+--+
| insert_data.name  |
+-------------------+--+
| Sophie            |
+-------------------+--+
1 row selected (0.327 seconds)
```

其中，执行插入操作时，Hive 会启动 MapReduce。

7.6.2 对表进行分区

Hive 中的分区是指根据表的列对表进行切分的机制。Hive 对表进行分区的目的是加快查询数据的速度。创建带分区的表使用 partitioned by 语句。本书将带分区的表称作分区表。

Hive 表的分区的添加方式有两种：静态添加分区和动态添加分区。

静态添加分区是在表中手动添加具有确定分区值的分区。Hive 表中默认使用的是静态添加分区。

动态添加分区是分区值不确定，根据输入数据自动添加分区。在实际开发工作中使用较多的是动态添加分区。

分区表的常见操作如下。

（1）创建分区表。

```
jdbc:hive2://> create table partition_tbl(id int, name string, age int)
partitioned by (province string, city string)
row format delimited
fields terminated by '\t'
lines terminated by '\n'
stored as textfile;
```

（2）在分区表中静态添加分区。

① 添加 1 个分区。

```
jdbc:hive2://> alter table partition_tbl add
partition(province='jiangsu', city='nanjing');
```

② 添加多个分区。

```
jdbc:hive2://> alter table partition_tbl add
partition(province='beijing', city='beijing')
partition(province='zhejiang', city='hangzhou')
partition(province='shandong', city='jinan');
```

（3）加载数据到分区。

① 准备数据。在虚拟机 qf01 的 /root 目录下，新建文件 partitiondata.txt，在文件中添加以下数据。注意：数据中的分隔符使用的是制表符 Tab 键。

```
1101	Sophie	30
1106	Coco	38
1108	Tom	27
```

② 加载 partitiondata.txt 的数据到表 partition_tbl 的指定分区。

```
jdbc:hive2://> load data local inpath '/root/partitiondata.txt' overwrite into table partition_tbl partition(province='jiangsu', city='nanjing');
```

（4）动态添加分区。一般在向表中插入数据或加载数据时进行动态添加分区。

① 查看当前 Hive 表是否可以执行动态添加分区。

```
jdbc:hive2://> set hive.exec.dynamic.partition;
+----------------------------------+--+
|               set                |
+----------------------------------+--+
| hive.exec.dynamic.partition=false |
+----------------------------------+--+
1 row selected (0.013 seconds)
```

② 设置可以执行动态添加分区。

```
jdbc:hive2://> set hive.exec.dynamic.partition=true;
```

③ 分区模式。Hive 的分区模式分为严格模式和非严格模式。严格模式要求 Hive 表的分区中至少有一个静态添加的分区。非严格模式下 Hive 表的分区均可以进行动态添加操作。

查看分区模式。

```
jdbc:hive2://> set hive.exec.dynamic.partition.mode;
```

```
+------------------------------------------+--+
|                   set                    |
+------------------------------------------+--+
| hive.exec.dynamic.partition.mode=strict  |
+------------------------------------------+--+
1 row selected (0.013 seconds)
```

设置动态添加分区的严格模式。

```
jdbc:hive2://> set hive.exec.dynamic.partition.mode=strict;
```

设置动态添加分区的非严格模式。

```
jdbc:hive2://> set hive.exec.dynamic.partition.mode=nonstrict;
```

④ 将未分区表中的数据动态插入分区表，并动态添加分区。

创建未分区表 nonpartition_tbl。

```
jdbc:hive2://> create table nonpartition_tbl(id int, name string, age int, province string, city string)
row format delimited
fields terminated by '\t'
lines terminated by '\n'
stored as textfile;
```

在虚拟机 qf01 的 /root 目录下新建文件 partdata.txt，添加如下数据（数据分隔符为制表符 Tab 键）。

```
1101    Sophie  30      beijing   beijing
1106    Coco    38      jiangsu   nanjing
1108    Tom     27      shandong  jinan
```

加载数据到表 nonpartition_tbl 中。

```
jdbc:hive2://> load data local inpath '/root/partdata.txt' into table nonpartition_tbl;
```

创建分区表 dynamic_partition_tbl。

```
jdbc:hive2://> create table dynamic_partition_tbl(id int, name string, age int)
partitioned by (province string, city string)
row format delimited
fields terminated by '\t'
lines terminated by '\n'
stored as textfile;
```

将表 nonpartition_tbl 中的数据动态插入表 dynamic_partition_tbl。

```
jdbc:hive2://> insert into table dynamic_partition_tbl partition(province, city)
select * from nonpartition_tbl;
```

查看表 dynamic_partition_tb 的数据。

```
jdbc:hive2://> select * from dynamic_partition_tbl;
+--------------------------+----------------------------+---------------------------
--+--------------------------+------------------------+--+
| dynamic_partition_tbl.id | dynamic_partition_tbl.name | dynamic_partition_tbl.age |
  dynamic_partition_tbl.province | dynamic_partition_tbl.city |
+--------------------------+----------------------------+---------------------------
--+--------------------------+------------------------+--+
| 1101                     | Sophie                     | 30                        | beijing       | beijing    |
| 1106                     | Coco                       | 38                        | jiangsu       | nanjing    |
```

```
| 1108                     | Tom                      | 27                       | shandong                 | jinan                    |
+--------------------------+--------------------------+--------------------------+--------------------------+--------------------------+
```

返回结果如上所示,表明动态数据插入成功。
查看表 dynamic_partition_tb 中动态添加的分区。

```
jdbc:hive2://> show partitions dynamic_partition_tbl;
+-------------------------------+--+
|           partition           |
+-------------------------------+--+
| province=beijing/city=beijing |
| province=jiangsu/city=nanjing |
| province=shandong/city=jinan  |
+-------------------------------+--+
3 rows selected (0.367 seconds)
```

(5)删除分区。

```
alter table partition_tbl drop if exists partition(province='beijing', city='beijing');
```

其中,if exists 是指如果存在指定的分区就删除该分区,是一个可选项。
(6)查看分区表中指定分区的数据。
查看分区表中指定分区的数据,主要有以下两种方式。
① 通过 select 查询语句来查看数据。

```
jdbc:hive2://> select * from partition_tbl where province='jiangsu' and city='nanjing';
+--------------------+----------------------+---------------------+--------------------------+--------------------+--+
| partition_tbl.id   | partition_tbl.name   | partition_tbl.age   | partition_tbl.province   | partition_tbl.city |
+--------------------+----------------------+---------------------+--------------------------+--------------------+--+
| 1101               | Sophie               | 30                  | jiangsu                  | nanjing            |
| 1106               | Coco                 | 38                  | jiangsu                  | nanjing            |
| 1108               | Tom                  | 27                  | jiangsu                  | nanjing            |
+--------------------+----------------------+---------------------+--------------------------+--------------------+--+
3 rows selected (0.298 seconds)
```

② 通过 WebUI 来查看数据。找到分区在 HDFS 上对应的目录。分区在 HDFS 中的存储形式是目录,可以在 HDFS 的/user/hive/warehouse/partition_tbl 目录下找到分区目录,如图 7.3 所示。

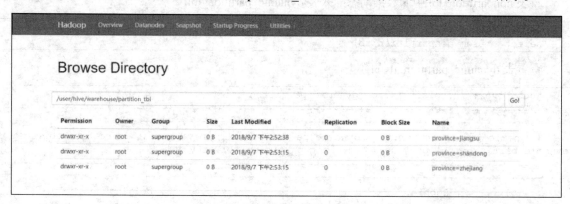

图 7.3 分区在 HDFS 中的存储形式

单击图 7.3 中的"province=jiangsu",依次在新打开的页面中单击"city=nanjing""partitiondata.txt"选项,进入分区中数据的下载界面,如图 7.4 所示。

图 7.4　分区中数据的下载界面

单击图 7.4 中的"Download"选项,下载指定分区的数据。可以使用记事本查看 partitiondata.txt 文件的数据。

提示:至此可知,Hive 的库、表、分区在 HDFS 中的存储形式都是目录,即 Hive 在创建库和表、为表添加分区时,都会在 HDFS 中新建目录或子目录。

(7) 重命名分区。

① 查看表 partition_tbl 的分区。

```
jdbc:hive2://> show partitions partition_tbl;
+--------------------------------+--+
|            partition           |
+--------------------------------+--+
| province=jiangsu/city=nanjing  |
| province=shandong/city=jinan   |
| province=zhejiang/city=hangzhou|
+--------------------------------+--+
3 rows selected (0.751 seconds)
```

② 重命名表 partition_tbl 的一个分区。

```
jdbc:hive2://> alter table partition_tbl partition(province='jiangsu', city='nanjing')
rename to partition(province='hainan', city='haikou');
```

③ 查看重命名后表 partition_tbl 的分区。

```
jdbc:hive2://> show partitions partition_tbl;
+--------------------------------+--+
|            partition           |
+--------------------------------+--+
| province=hainan/city=haikou    |
| province=shandong/city=jinan   |
| province=zhejiang/city=hangzhou|
+--------------------------------+--+
3 rows selected (0.212 seconds)
```

7.6.3　对表或分区进行桶操作

Hive 的每个表或分区可以划分成桶(Bucket)。本书将划分了桶的表称为桶表。桶表在 HDFS 中的存储形式是文件。Hive 在启动 MapReduce 进行计算时产生桶的个数（输出文件数）和 Reduce 任务数相同。

Hive 将表或分区划分成桶主要有以下两个目的。

（1）当两个按相同列划分了桶的表进行连接（Join）操作时，可以减少 Join 的数据量，提高连接效率。

（2）在处理大数据集时，对划分了桶的数据进行采样处理，可以提高采样的效率。

Hive 中使用 clustered by 语句来对表或分区进行桶操作。

下面通过示例来理解桶操作。

1. 创建桶表并插入数据

（1）创建桶表。

```
jdbc:hive2://> create table bucket_tbl(id int, name string, age int, province string, city string)
clustered by(province) into 2 buckets
row format delimited
fields terminated by '\t';
```

（2）设置 Hive 可以执行桶操作。

```
jdbc:hive2://> set hive.enforce.bucketing=true;
```

（3）向桶表中插入数据。

```
jdbc:hive2://> insert into bucket_tbl select * from nonpartition_tbl;
```

（4）查看桶表的数据。

```
jdbc:hive2://> select * from bucket_tbl;
+----------------+------------------+----------------+--------------------+------------------+
| bucket_tbl.id  | bucket_tbl.name  | bucket_tbl.age | bucket_tbl.province| bucket_tbl.city  |
+----------------+------------------+----------------+--------------------+------------------+
| 1108           | Tom              | 27             | shandong           | jinan            |
| 1101           | Sophie           | 30             | beijing            | beijing          |
| 1106           | Coco             | 38             | jiangsu            | nanjing          |
+----------------+------------------+----------------+--------------------+------------------+
3 rows selected (0.545 seconds)
```

（5）通过 WebUI 查看桶表 bucket_tbl 的数据。

打开 http://192.168.142.131:50070/explorer.html#/user/hive/warehouse/bucket_tbl，查看桶表 bucket_tbl 在 HDFS 上对应的文件，如图 7.5 所示。

图 7.5　桶表 bucket_tbl 在 HDFS 上对应的文件

由图 7.5 可知，桶表 bucket_tbl 在 HDFS 上对应的文件有两个：000000_0 和 000001_0。下载这两个文件，用记事本分别查看里面的数据。

000000_0 文件的数据如下。

```
1108 Tom    27   shandong jinan
1101 Sophie 30   beijing  beijing
```

000001_0 文件的数据如下。

```
1106 Coco 38    jiangsu  nanjing
```

2. 创建带分区的桶表（外部表）并插入数据

（1）创建带分区的桶表。

```
jdbc:hive2://> create external table bucket_partition_extbl(id int, name string, age int)
partitioned by(province string , city string)
clustered by(id) into 2 buckets
row format delimited
fields terminated by '\t';
```

（2）设置 Hive 可以执行动态添加分区和 Hive 动态添加分区的非严格模式。

```
jdbc:hive2://> set hive.exec.dynamic.partition=true;
jdbc:hive2://> set hive.exec.dynamic.partition.mode=nonstrict;
```

（3）向分区中插入数据。

```
jdbc:hive2://> insert into bucket_partition_extbl partition(province, city) select * from nonpartition_tbl;
```

（4）查看数据是否插入成功。

```
jdbc:hive2://> select * from bucket_partition_extbl;
+--------------------------+---------------------------+--------------------------+------------------------------+---------------------------+--+
| bucket_partition_extbl.id | bucket_partition_extbl.name | bucket_partition_extbl.age | bucket_partition_extbl.province | bucket_partition_extbl.city |
+--------------------------+---------------------------+--------------------------+------------------------------+---------------------------+--+
| 1101                     | Sophie                    | 30                       | beijing                      | beijing                   |
| 1106                     | Coco                      | 38                       | jiangsu                      | nanjing                   |
| 1108                     | Tom                       | 27                       | shandong                     | jinan                     |
+--------------------------+---------------------------+--------------------------+------------------------------+---------------------------+--+
3 rows selected (2.751 seconds)
```

（5）通过 WebUI 查看插入桶表的数据在 HDFS 上的存放位置。

在 HDFS 的/user/hive/warehouse/bucket_partition_extbl 目录下，可以发现有 3 个分区，在每个分区下各存放了两个桶表文件：000000_0 和 000001_0。表 bucket_partition_extbl 中的 3 条数据随机存储在如下 3 个位置。

```
province=beijing/city=beijing 下的 000001_0 文件。
province=jiangsu/city=nanjing 下的 000000_0 文件。
province=shandong/city=jinan 下的 000000_0 文件。
```

（6）向桶表中插入多条数据。

```
jdbc:hive2://> insert into table bucket_partition_extbl partition (province='henan',
```

```
city='zhengzhou') select id, name, age from nonpartition_tbl;
    jdbc:hive2://> select * from bucket_partition_extbl ;
    +---------------------------+-----------------------------+----------------------------+
-----+----------------------------+-----------------------------+--+
    | bucket_partition_extbl.id | bucket_partition_extbl.name | bucket_partition_extbl.age
 | bucket_partition_extbl.province | bucket_partition_extbl.city    |
    +---------------------------+-----------------------------+----------------------------+
-----+----------------------------+-----------------------------+--+
    | 1101                      | Sophie                      | 30                         | beijing                    | beijing                     |
    | 1108                      | Tom                         | 27                         | henan                      | zhengzhou                   |
    | 1106                      | Coco                        | 38                         | henan                      | zhengzhou                   |
    | 1101                      | Sophie                      | 30                         | henan                      | zhengzhou                   |
    | 1106                      | Coco                        | 38                         | jiangsu                    | nanjing                     |
    | 1108                      | Tom                         | 27                         | shandong                   | jinan                       |
    +---------------------------+-----------------------------+----------------------------+
-----+----------------------------+-----------------------------+--+
    6 rows selected (0.498 seconds)
```

查看 WebUI 可知，新插入的 3 条数据存储在新添加的分区目录 province=henan/city=zhengzhou 下的两个文件中。

（7）查看桶表的采样（Sampling）数据，采样的规则是在两个桶中随机选取 1 个桶的数据。

```
    jdbc:hive2://> select * from bucket_partition_extbl tablesample(bucket 1 out of 2 on id);
    +---------------------------+-----------------------------+----------------------------+
-----+----------------------------+-----------------------------+--+
    | bucket_partition_extbl.id | bucket_partition_extbl.name | bucket_partition_extbl.age
 | bucket_partition_extbl.province | bucket_partition_extbl.city    |
    +---------------------------+-----------------------------+----------------------------+
-----+----------------------------+-----------------------------+--+
    | 1108                      | Tom                         | 27                         | henan                      | zhengzhou                   |
    | 1106                      | Coco                        | 38                         | henan                      | zhengzhou                   |
    | 1106                      | Coco                        | 38                         | jiangsu                    | nanjing                     |
    | 1108                      | Tom                         | 27                         | shandong                   | jinan                       |
    +---------------------------+-----------------------------+----------------------------+
-----+----------------------------+-----------------------------+--+
    4 rows selected (0.911 seconds)
```

其中，tablesample 是用于采样的关键字，语法如下。

```
tablesample(bucket x out of y)
```

bucket x out of y 是指从 y 个桶里选取 x 个桶的数据。参数 y 往往是 x（x≤y）的正整数倍。

7.7 Hive 表的查询

7.7.1 select 查询语句

1. select 查询语句的语法

Hive 表的查询操作主要是通过 select 查询语句来实现的。select 查询语句的基本语法如下。

```
SELECT [ALL | DISTINCT] select_expr, select_expr, ...
FROM table_reference
[WHERE where_condition]
[GROUP BY col_list [HAVING condition]]
[CLUSTER BY col_list
```

```
   | [DISTRIBUTE BY col_list] [SORT BY | ORDER BY col_list]
   ]
   [LIMIT number]
```

2. select 查询语句解析及示例

（1）select 和 from。select 是查询语句的关键字。from 用于定位查询的表。

查看表中指定列的数据。

```
jdbc:hive2://> alter table bucket_partition_extbl rename to select_tbl;
jdbc:hive2://> select id, name from select_tbl;
+-------+---------+--+
|  id   |  name   |
+-------+---------+--+
| 1101  | Sophie  |
| 1108  | Tom     |
| 1106  | Coco    |
| 1101  | Sophie  |
| 1106  | Coco    |
| 1108  | Tom     |
+-------+---------+--+
6 rows selected (0.284 seconds)
```

（2）all 和 distinct。all 用于查询所有的数据。distinct 用于对查询的数据去重。

查看表中指定列的所有数据。

```
jdbc:hive2://> select all name, age from select_tbl;
+---------+------+--+
|  name   | age  |
+---------+------+--+
| Sophie  | 30   |
| Tom     | 27   |
| Coco    | 38   |
| Sophie  | 30   |
| Coco    | 38   |
| Tom     | 27   |
+---------+------+--+
6 rows selected (0.297 seconds)
```

查看表中指定列去重后的数据。

```
jdbc:hive2://> select distinct name, age from select_tbl;
+---------+------+--+
|  name   | age  |
+---------+------+--+
| Coco    | 38   |
| Sophie  | 30   |
| Tom     | 27   |
+---------+------+--+
3 rows selected (38.972 seconds)
```

使用 distinct 关键字时，需要启动 MapReduce 作业，可以通过 Hiveserver2 服务端界面查看作业运行信息。其他操作同样可以通过 Hiveserver2 查看相关信息，不再赘述。

（3）where。where 用于设置过滤条件。

```
jdbc:hive2://> select distinct id, name, age from select_tbl where city ='zhengzhou';
+-------+---------+------+--+
```

```
| id   | name    | age |
+------+---------+-----+
| 1101 | Sophie  | 30  |
| 1106 | Coco    | 38  |
| 1108 | Tom     | 27  |
+------+---------+-----+
3 rows selected (33.128 seconds)
```

(4) group by。group by 用于根据指定条件对数据进行分组去重查询。group by 常与 Hive 函数（如 count()等）一起使用。

```
jdbc:hive2://> select name, count(city) as citys from select_tbl group by name;
+--------+-------+
| name   | citys |
+--------+-------+
| Coco   | 2     |
| Sophie | 2     |
| Tom    | 2     |
+--------+-------+
3 rows selected (26.797 seconds)
```

其中，as citys 是指为新的临时列取别名为 citys。

(5) having。having 用于对每个组内的数据进行过滤。

```
jdbc:hive2://> select name, count(city) as citys from select_tbl group by name;
+--------+-------+
| name   | citys |
+--------+-------+
| Sophie | 2     |
+--------+-------+
1 row selected (59.542 seconds)
```

(6) limit。limit 用于限制查询。

查看表 select_tbl 的前 5 行数据。

```
jdbc:hive2://> select * from select_tbl limit 5;
+----------------+------------------+-----------------+----------------------+------------------+
| select_tbl.id  | select_tbl.name  | select_tbl.age  | select_tbl.province  | select_tbl.city  |
+----------------+------------------+-----------------+----------------------+------------------+
| 1101           | Sophie           | 30              | beijing              | beijing          |
| 1108           | Tom              | 27              | henan                | zhengzhou        |
| 1106           | Coco             | 38              | henan                | zhengzhou        |
| 1101           | Sophie           | 30              | henan                | zhengzhou        |
| 1106           | Coco             | 38              | jiangsu              | nanjing          |
+----------------+------------------+-----------------+----------------------+------------------+
5 rows selected (1.679 seconds)
```

查看表 select_tbl 第 3 行后的 3 行数据（不包括第 3 行的数据）。

```
jdbc:hive2://> select * from select_tbl limit 3,3;
+----------------+------------------+-----------------+----------------------+------------------+
| select_tbl.id  | select_tbl.name  | select_tbl.age  | select_tbl.province  | select_tbl.city  |
```

```
+----------------+----------------+----------------+-------------------+----------------+--+
| 1101           | Sophie         | 30             | henan             | zhengzhou      |
| 1106           | Coco           | 38             | jiangsu           | nanjing        |
| 1108           | Tom            | 27             | shandong          | jinan          |
+----------------+----------------+----------------+-------------------+----------------+--+
3 rows selected (0.282 seconds)
```

（7）order by。order by 用于对所有数据排序。Hive 中使用 order by 会使所有的数据都通过一个 Reduce 任务来处理（多个 Reduce 任务无法保证全局有序），如果处理的数据量较大，则需要耗费较长的计算时间，因此，在使用 order by 时，为了优化查询的速度，需要设置 Hive 执行 MapReduce 的模式为严格模式，同时指定 limit 来限制输出的数据量。

设置 Hive 执行 MapReduce 的模式为严格模式（默认是非严格模式）。

```
jdbc:hive2://> set hive.mapred.mode=strict;
```

设置严格模式下检查大数据量的查询。

```
jdbc:hive2://> hive.strict.checks.large.query=true;
```

对表 select_tbl 的前 5 行进行排序。

```
jdbc:hive2://> select * from select_tbl order by id limit 5;
+----------------+-------------------+-----------------+----------------------+----------------+--+
| select_tbl.id  | select_tbl.name   | select_tbl.age  | select_tbl.province  | select_tbl.city |
+----------------+-------------------+-----------------+----------------------+----------------+--+
| 1101           | Sophie            | 30              | henan                | zhengzhou      |
| 1101           | Sophie            | 30              | beijing              | beijing        |
| 1106           | Coco              | 38              | jiangsu              | nanjing        |
| 1106           | Coco              | 38              | henan                | zhengzhou      |
| 1108           | Tom               | 27              | shandong             | jinan          |
+----------------+-------------------+-----------------+----------------------+----------------+--+
5 rows selected (34.82 seconds)
```

（8）sort by。sort by 用于对同一个 Reduce 任务中的数据进行排序。与 order by 相比，sort by 不受 hive.mapred.mode 的影响。

sort by 排序示例。

```
jdbc:hive2://> select name, age from select_tbl sort by age;
+---------+------+--+
| name    | age  |
+---------+------+--+
| Tom     | 27   |
| Tom     | 27   |
| Sophie  | 30   |
| Sophie  | 30   |
| Coco    | 38   |
| Coco    | 38   |
+---------+------+--+
6 rows selected (37.97 seconds)
```

sort by 倒排序（desc）示例。

```
jdbc:hive2://> select name, age from select_tbl sort by age desc;
+---------+------+--+
|  name   | age  |
+---------+------+--+
| Coco    | 38   |
| Coco    | 38   |
| Sophie  | 30   |
| Sophie  | 30   |
| Tom     | 27   |
| Tom     | 27   |
+---------+------+--+
6 rows selected (56.799 seconds)
```

在使用 sort by 时,可以指定执行 MapReduce 作业的 Reduce 个数,具体格式如下。

```
jdbc:hive2://> set mapreduce.job.reduces=<number>;
```

示例如下。

```
jdbc:hive2://> set mapreduce.job.reduces=2;
jdbc:hive2://> select name, age from select_tbl sort by age;
+---------+------+--+
|  name   | age  |
+---------+------+--+
| Sophie  | 30   |
| Sophie  | 30   |
| Coco    | 38   |
| Coco    | 38   |
| Tom     | 27   |
| Tom     | 27   |
+---------+------+--+
6 rows selected (52.014 seconds)
```

查看 Hiveserver2 服务端,可以发现执行了 2 个 Reduce 任务,摘录部分信息如下。

```
MapReduce Jobs Launched:
Stage-Stage-1: Map: 1  Reduce: 2
```

(9) distribute by。distribute by 用于按照列使数据分散到不同的 Reduce 任务中。distribute by 相当于 Hadoop 中 MapReduce 的分区操作,与 sort by 组合使用时,必须放在 sort by 语句之前。

单独使用 distribute by 的示例。

```
jdbc:hive2://> set mapreduce.job.reduces=3;
jdbc:hive2://> select id, name from select_tbl distribute by id;
+-------+---------+--+
|  id   |  name   |
+-------+---------+--+
| 1101  | Sophie  |
| 1101  | Sophie  |
| 1108  | Tom     |
| 1108  | Tom     |
| 1106  | Coco    |
| 1106  | Coco    |
+-------+---------+--+
6 rows selected (124.065 seconds)
```

distribute by 和 sort by 同时使用的示例。

```
jdbc:hive2://> select name, age from select_tbl distribute by name sort by age desc;
+--------+------+--+
| name   | age  |
+--------+------+--+
| Coco   | 38   |
| Coco   | 38   |
| Tom    | 27   |
| Tom    | 27   |
| Sophie | 30   |
| Sophie | 30   |
+--------+------+--+
6 rows selected (49.774 seconds)
```

由查询结果可知，6 行数据被分配到 3 个 Reduce 任务中进行了处理。

（10）cluster by。cluster by 相当于使用相同列的 distribute by 和 sort by 的结合，默认只能对数据进行升序排列，不支持倒排序。

例如，使用 1 个 Reduce 实现数据的全排序。

```
jdbc:hive2://> set mapreduce.job.reduces=1;
jdbc:hive2://> select id, name from select_tbl cluster by id;
jdbc:hive2://> select id, name from select_tbl distribute by id sort by id;
+-------+---------+--+
| id    | name    |
+-------+---------+--+
| 1101  | Sophie  |
| 1101  | Sophie  |
| 1106  | Coco    |
| 1106  | Coco    |
| 1108  | Tom     |
| 1108  | Tom     |
+-------+---------+--+
6 rows selected (31.638 seconds)
```

由查询结果可知，以上两种查询语句的作用是相同的。

（11）子查询。子查询又称嵌套查询，是多个 select 语句嵌套的查询方式。

```
jdbc:hive2://> select distinct a.id, a.name from (select * from select_tbl limit 5) a;
+-------+---------+--+
| a.id  | a.name  |
+-------+---------+--+
| 1101  | Sophie  |
| 1106  | Coco    |
| 1108  | Tom     |
+-------+---------+--+
3 rows selected (85.322 seconds)
```

其中，a 是内层 select 查询语句所生成临时表的别名。

7.7.2 视图

Hive 中的视图（View）是一种不存放真实数据的虚表。视图可以用于简化数据查询语句。视图是只读的，不能向视图中插入或是加载数据，一定程度上可以提高数据的安全性。视图的常用操作如下所示。

(1) 创建视图。

```
jdbc:hive2://> create view view_temp as select id, name, age from select_tbl;
```

创建视图后,查看 HDFS 的 /user/hive/warehouse 目录下的内容可知,视图并没有在 HDFS 中存放真实的数据。

(2) 查看视图的结构。

```
jdbc:hive2://> desc view_temp;
+-----------+-----------+----------+--+
| col_name  | data_type | comment  |
+-----------+-----------+----------+--+
| id        | int       |          |
| name      | string    |          |
| age       | int       |          |
+-----------+-----------+----------+--+
3 rows selected (0.093 seconds)
```

(3) 查看视图的数据。

```
jdbc:hive2://> select * from view_temp;
+----------------+------------------+-----------------+--+
| view_temp.id   | view_temp.name   | view_temp.age   |
+----------------+------------------+-----------------+--+
| 1101           | Sophie           | 30              |
| 1108           | Tom              | 27              |
| 1106           | Coco             | 38              |
| 1101           | Sophie           | 30              |
| 1106           | Coco             | 38              |
| 1108           | Tom              | 27              |
+----------------+------------------+-----------------+--+
6 rows selected (0.658 seconds)
```

当使用 select 查询语句查看视图的数据时,实际上是引用了创建视图时的 select id, name, age from select_tbl select 语句的查询结果。

(4) 删除视图。

```
jdbc:hive2://> drop view if exists view_temp;
```

7.7.3 Join

Hive 的 Join 操作主要有内连接、外连接、半连接。下面对 Join 操作进行演示。

1. 准备数据

创建两个表,并导入数据:表 j1 存储学生的姓名和学号;表 j2 存储学生的学号和籍贯所在的地级市。

(1) 创建文件。在虚拟机 qf01 的 /root 目录下新建两个文件:join1.txt 和 join2.txt。
join1.txt 文件的数据如下(数据间的分隔符是制表符)。

```
Sophie    3
Coco      6
Tom       5
Jack      2
Rose      1
```

join2.txt 文件的数据如下（数据间的分隔符是制表符）。

```
3       beijing
6       tianjin
5       shanghai
8       hangzhou
9       chengdu
2       nanjing
```

（2）创建表 j1 和表 j2。

```
jdbc:hive2://> create table j1(name string, id int)
row format delimited
fields terminated by '\t'
lines terminated by '\n'
stored as textfile;
jdbc:hive2://> create table j2(id int, city string)
row format delimited
fields terminated by '\t'
lines terminated by '\n'
stored as textfile;
```

（3）加载数据到表中。分别加载虚拟机 qf01 的/root 目录下的 join1.txt 和 join2.txt 文件到表 j1 和表 j2 中。

```
jdbc:hive2://> load data local inpath '/root/join1.txt' into table j1;
jdbc:hive2://> load data local inpath '/root/join2.txt' into table j2;
```

2. 内连接

内连接相当于取两个表中数据的交集。

```
jdbc:hive2://> select j1.*, j2.* from j1 join j2 on (j1.id=j2.id);
+---------+-------+-------+----------+--+
| j1.name | j1.id | j2.id | j2.city  |
+---------+-------+-------+----------+--+
| Sophie  | 3     | 3     | beijing  |
| Coco    | 6     | 6     | tianjin  |
| Tom     | 5     | 5     | shanghai |
| Jack    | 2     | 2     | nanjing  |
+---------+-------+-------+----------+--+
4 rows selected (55.8 seconds)
```

该语句与下面的语句的作用是相同的。

```
jdbc:hive2://> select j1.*, j2.* from j1, j2 where (j1.id=j2.id);
```

注意：Hive 只支持等值连接，例如，j1.id 和 j2.id 之间只能使用等号。

3. 外连接

外连接分为左外连接（left outer join）、右外连接（right outer join）、全外连接（full outer join）。

（1）左外连接。左外连接相当于左侧表的所有数据与两个表的交集数据取并集。

```
jdbc:hive2://> select j1.*, j2.* from j1 left outer join j2 on (j1.id=j2.id);
+---------+-------+-------+----------+--+
| j1.name | j1.id | j2.id | j2.city  |
+---------+-------+-------+----------+--+
| Sophie  | 3     | 3     | beijing  |
| Coco    | 6     | 6     | tianjin  |
```

```
| Tom    | 5    | 5    | shanghai |
| Jack   | 2    | 2    | nanjing  |
| Rose   | 1    | NULL | NULL     |
+--------+------+------+----------+--+
5 rows selected (34.617 seconds)
```

由查询结果可知,Rose 一行的 j2.id 和 j2.city 两列的数据为空值 NULL,原因在于 j1.id 在表 j2 中没有匹配项。

(2)右外连接。右外连接相当于右侧表的所有数据与两个表的交集数据取并集。

```
jdbc:hive2://> select j1.*, j2.* from j1 right outer join j2 on (j1.id=j2.id);
+---------+-------+-------+----------+--+
| j1.name | j1.id | j2.id | j2.city  |
+---------+-------+-------+----------+--+
| Sophie  | 3     | 3     | beijing  |
| Coco    | 6     | 6     | tianjin  |
| Tom     | 5     | 5     | shanghai |
| NULL    | NULL  | 8     | hangzhou |
| NULL    | NULL  | 9     | chengdu  |
| Jack    | 2     | 2     | nanjing  |
+---------+-------+-------+----------+--+
6 rows selected (33.568 seconds)
```

由查询结果可知,第 4 行和第 5 行的 j1.name 和 j1.id 两列的数据为空值 NULL,原因在于 j2.id 在表 j1 中没有匹配项。

(3)全外连接。全外连接相当于两个表的所有数据取并集。

```
jdbc:hive2://> select j1.*, j2.* from j1 full outer join j2 on (j1.id=j2.id);
+---------+-------+-------+----------+--+
| j1.name | j1.id | j2.id | j2.city  |
+---------+-------+-------+----------+--+
| Rose    | 1     | NULL  | NULL     |
| Jack    | 2     | 2     | nanjing  |
| Sophie  | 3     | 3     | beijing  |
| Tom     | 5     | 5     | shanghai |
| Coco    | 6     | 6     | tianjin  |
| NULL    | NULL  | 8     | hangzhou |
| NULL    | NULL  | 9     | chengdu  |
+---------+-------+-------+----------+--+
7 rows selected (44.669 seconds)
```

4. 半连接

半连接相当于两个表的数据取交集,只显示左表的指定数据。

```
jdbc:hive2://> select city from j2 left semi join j1 on (j1.id=j2.id);
+----------+--+
| city     |
+----------+--+
| beijing  |
| tianjin  |
| shanghai |
| nanjing  |
+----------+--+
4 rows selected (32.219 seconds)
```

注意:所有 Join 操作的顺序都是从左到右,查询语句中左侧的表为左表,右侧的表为右表。

5. Map Join

Map Join 是将较小的表加载到 Mapper 内存来执行的连接操作。Hive 只执行 Map Join 而不执行 Reduce 时，一定程度上可以节省计算资源。查看 Hive 是否启用了 Map Join 的方法如下（Hive 默认启用 Map Join）。

```
jdbc:hive2://> set hive.auto.convert.join;
```

返回 true，表明 Hive 使用了 Map Join。

例如，内连接示例就使用了 Map Join 操作。

```
jdbc:hive2://> select j1.*, j2.* from j1 join j2 on (j1.id=j2.id);
```

执行以上查询语句，查看 Hiveserver2 服务端，可以发现如下信息（节选）。

```
2018-09-14 11:20:31,186 Stage-3 map = 100%,  reduce = 0%, Cumulative CPU 6.36 sec
MapReduce Total cumulative CPU time: 6 seconds 360 msec
Ended Job = job_1536887505462_0016
MapReduce Jobs Launched:
Stage-Stage-3: Map: 1
```

由以上信息可知，内连接的操作不需要执行 Reduce 阶段。

6. Join 和桶表

我们已经知道，Hive 对表或分区划分桶的目的之一是：当两个按相同列划分了桶的表进行连接（Join）操作时，可以减少 Join 的数据量，提高连接效率。

（1）创建两个桶表。

```
jdbc:hive2://> create table join_bucket1(id int, name string, age int, province string, city string)
clustered by(id) into 2 buckets
row format delimited
fields terminated by '\t';
jdbc:hive2://> create table join_bucket2(id int, name string, age int, province string, city string)
clustered by(id) into 2 buckets
row format delimited
fields terminated by '\t';
```

（2）向桶表中分别插入数据。

```
jdbc:hive2://> insert into join_bucket1 select * from nonpartition_tbl;
jdbc:hive2://> insert into join_bucket2 select * from select_tbl;
```

（3）表 join_bucket1 和表 join_bucket2 进行内连接操作。

```
jdbc:hive2://> select join_bucket1.id, join_bucket1.name, join_bucket2.id, join_bucket2.name from join_bucket1 join join_bucket2 on (join_bucket1.id= join_bucket2.id);
```

查看 Hiveserver2 服务端的信息可知，表 join_bucket1 和 join_bucket2 进行内连接操作时，只使用 Map Join 就完成了内连接操作，未启用 Reduce 任务，减少了 Join 的数据量，提高了查询数据的效率。

7.8 Hive 函数

7.8.1 Hive 内置函数

为了方便大数据开发人员使用 Hive 查询和分析数据，Hive 提供了一些内置函数。常用的 Hive

内置函数有 3 类：内置普通函数、内置聚合函数、内置表生成函数。

1. 常用内置普通函数

（1）查看 Hive 的所有函数。

```
jdbc:hive2://> show functions;
```

（2）查看指定函数的用法，以 substr()函数为例。

```
jdbc:hive2://> desc function substr;
```

（3）查看指定函数的详细用法，以 substr()函数为例。部分函数提供范例。

```
jdbc:hive2://> desc function extended substr;
```

（4）显示当前日期。

```
jdbc:hive2://> select current_date();
```

（5）显示详细时间。

```
jdbc:hive2://> select current_timestamp();
```

（6）转换日期格式。

```
jdbc:hive2://> select date_format('2018-9-20','yyyy/MM/dd');
```

2. 常用内置聚合函数

聚合函数是将多行数据按照指定规则聚集为单行数据的函数。常用的内置聚合函数有 count()、sum()、avg()、max()、min()等。下面通过示例来理解聚合函数的用法。

根据表 select_tbl 进行人数统计，计算年龄平均值，查看年龄最大值、年龄最小值。具体示例如下。

```
jdbc:hive2://> select count(name), sum(age), cast(avg(age) as float), max(age), min(age) from select_tbl;
+-----+------+------------+-----+-----+--+
| c0  | c1   |    c2      | c3  | c4  |
+-----+------+------------+-----+-----+--+
| 6   | 190  | 31.666666  | 38  | 27  |
+-----+------+------------+-----+-----+--+
1 row selected (44.041 seconds)
```

其中，cast()是类型转换函数，用来将平均值计算结果转换为浮点数。

3. 常用内置表生成函数

表生成函数是将单行数据按照指定规则转换为多行数据的函数。常用的内置表生成函数有 explode(array)、explode(map)。

explode(array)将数组的元素分解为多行单列的表，每个元素各占一行。

explode(map)将 map 的键值对分解为多行两列的表，键、值各占一列。

下面通过示例来理解这两个表生成函数的用法。

（1）explode(array)示例如下。

```
jdbc:hive2://> select explode(array('a','b','c'));
+------+--+
| col  |
+------+--+
| a    |
| b    |
```

```
| c    |
+------+--+
3 rows selected (0.183 seconds)
```

(2) explode(map)示例如下。

```
jdbc:hive2://> select explode(map(1,'hello', 2,'world', 3,'hadoop'));
+------+---------+--+
| key  |  value  |
+------+---------+--+
| 1    | hello   |
| 2    | world   |
| 3    | hadoop  |
+------+---------+--+
3 rows selected (0.215 seconds)
```

4. 使用 Hive 函数实现 WordCount

(1) 创建表 wordcount。

```
jdbc:hive2://> create table wordcount(line string)
row format delimited
fields terminated by '\t'
lines terminated by '\n'
stored as textfile;
```

(2) 加载 HDFS 的 /word.txt 文件到 wordcount 表。

```
jdbc:hive2://> load data inpath '/word.txt' into table wordcount;
```

(3) 实现 WordCount。

```
jdbc:hive2://> select a.word word, count(*) b
from(
select explode(split(line, ' ')) as word
from wordcount
) a
group by a.word
order by b desc
limit 6;
+------------+----+--+
|    word    | b  |
+------------+----+--+
| hi         | 3  |
| qianfeng   | 2  |
| hello      | 2  |
| world      | 1  |
| mapreduce  | 1  |
| hadoop     | 1  |
+------------+----+--+
6 rows selected (90.772 seconds)
```

显然，对于实现简单的 WordCount 而言，使用 Hive 函数比使用 MapReduce 编程更加简便。

7.8.2　通过 JDBC 驱动程序使用 Hiveserver2 服务

通过 JDBC 驱动程序使用 Hiveserver2 服务的步骤如下。

1. 配置开发环境

在 IntelliJ IDEA 中项目 testHadoop 下新建模块 testHive,修改 pom.xml 文件内容如下。

```xml
<?xml version="1.0" encoding="UTF-8"?>
<project xmlns="http://maven.apache.org/POM/4.0.0"
         xmlns:xsi="http://www.w3.org/2001/XMLSchema-instance"
         xsi:schemaLocation="http://maven.apache.org/POM/4.0.0 http://maven.apache.org/xsd/maven-4.0.0.xsd">
    <modelVersion>4.0.0</modelVersion>
    <groupId>com.qf</groupId>
    <artifactId>testHive</artifactId>
    <version>1.0-SNAPSHOT</version>
    <dependencies>
        <dependency>
            <groupId>org.apache.hive</groupId>
            <artifactId>hive-jdbc</artifactId>
            <version>2.1.1</version>
        </dependency>
        <dependency>
            <groupId>org.apache.hive</groupId>
            <artifactId>hive-exec</artifactId>
            <version>2.1.1</version>
        </dependency>
        <dependency>
            <groupId>junit</groupId>
            <artifactId>junit</artifactId>
            <version>4.12</version>
        </dependency>
    </dependencies>
</project>
```

2. 编程

在 IntelliJ IDEA 中 testHive 模块的 src/main/java 文件夹下,新建类 com.qf.hive.TestJDBC。

【例 7-1】TestJDBC.java

```
1   package com.qf.hive;
2   import java.sql.Connection;
3   import java.sql.DriverManager;
4   import java.sql.ResultSet;
5   import java.sql.Statement;
6   public class TestJDBC {
7       public static void main(String[] args) throws Exception {
8           //1.设置连接地址(Hive 默认端口号为 10000,数据库为 default)
9           String url = "jdbc:hive2://192.168.142.131:10000/default";
10          //2.通过 url 获取连接
11          Connection conn = DriverManager.getConnection(url);
12          //3.通过连接创建语句
13          Statement st = conn.createStatement();
14          //4.执行 HQL 查询语句获取结果
15          ResultSet rs = st.executeQuery("select id, name from bucket_tbl");
16          //5.迭代输出查询结果
17          while (rs.next()) {
18              int id = rs.getInt(1);
19              String name = rs.getString(2);
```

```
20              System.out.println(id + "\t" + name + "\t");
21         }
22      }
23 }
```

3. 启动 Hiveserver2 服务

在虚拟机 qf01 上启动 Hive 的 Hiveserver2 服务。

```
[root@qf01 ~]# hiveserver2
```

4. 运行程序

运行例 7-1 的程序，在 IntelliJ IDEA 的控制台中，出现以下内容表明操作正确。

```
1108 Tom
1101 Sophie
1106 Cocona
```

7.8.3 Hive 用户自定义函数

Hive 的用户自定义函数主要有 3 种：用户自定义普通函数（User Defined Function，UDF）、用户自定义聚合函数（User-Defined Aggregate Funcation，UDAF）、用户自定义表生成函数（User-Defined Table-Generating Function，UDTF）。

由于 Hive 是使用 Java 语言编写的，因此 Hive 用户自定义函数也需要使用 Java 语言。下面以 UDF 为例，讲解 Hive 用户自定义函数的步骤。

设计实现加法的 UDF。通过该 UDF 实现：整数类型的数据进行累加，字符串类型的数据进行拼接，整数类型和字符串类型的数据进行拼接。

（1）编程实现 UDF。

```
1  package com.qf.hive;
2  import org.apache.hadoop.hive.ql.exec.Description;
3  import org.apache.hadoop.hive.ql.exec.UDF;
4  import java.util.ArrayList;
5  import java.util.List;
6  @Description(
7      name = "addition",    //函数名称
8      value = "this is an addition function.",   //函数作用描述
9      extended = "example:" +
10         "select addition(3,5,7) => 15 ;" +
11         " select addition('hi','xiaoqian') => hixiaoqian"   //函数应用举例
12 )
13 public class MyUDF extends UDF {
14     public Integer evaluate(int i, int j) {
15         return i + j;
16     }
17     public String evaluate(String i, String j) {
18         return i + j;
19     }
20     public String evaluate(ArrayList<String> i) {
21         String str = "";
22         for (int j = 0; j < i.size(); j++) {
23             str += i.get(j);
24         }
```

```
25          return str;
26      }
27      public int evaluate(List<Integer> i){
28          int j = 0;
29          for (Integer integer : i) {
30              j += integer;
31          }
32          return j;
33      }
34  }
```

编程时需要注意以下两点。
- 用户自定义普通函数必须继承 org.apache.hadoop.hive.ql.exec.UDF 类。
- 用户自定义普通函数必须含有 evaluate()方法，用户可以自定义 evaluate()方法的参数类型、参数个数、返回值类型。

（2）用 Maven Projects 将模块 testHive 打成 jar 包（testHive-1.0-SNAPSHOT.jar）。

（3）将 jar 包复制到虚拟机 qf01 的/mysoft/hive/lib 目录下，在 Beeline 客户端使用 jar 包。

```
jdbc:hive2://> add jar /mysoft/hive/lib/testHive-1.0-SNAPSHOT.jar;
```

（4）将 Java 类加载为 Hive 函数。

将 Java 类加载为 Hive 函数主要有以下两种方式。

创建临时函数。

```
jdbc:hive2://> create temporary function addition as 'com.qf.hive.MyUDF';
```

创建永久函数。

```
jdbc:hive2://> create function addition as 'com.qf.hive.MyUDF';
```

以下讲解以创建临时函数为例。

（5）查看 Hive 内置函数，出现 addition()函数，表明函数创建成功。

```
jdbc:hive2://> show functions;
```

（6）查看 addition()函数的详细用法。

```
jdbc:hive2://> desc function extended addition;
+---------------+--+
|    tab_name   |
+---------------+--+
| this is an addition function.            |
| example:select addition(3, 5, 7) => 15, select addition('hi', 'xiaoqian') => hixiaoqian |
+---------------+--+
2 rows selected (0.035 seconds)
```

（7）使用 addition()函数。

整数类型的数据进行累加。

```
jdbc:hive2://> select addition(array(1, 2, 3));
+------+--+
| _c0  |
+------+--+
| 6    |
+------+--+
```

```
1 row selected (0.741 seconds)
```

字符串类型的数据进行拼接。

```
jdbc:hive2://> select addition(array('hello', 'world'));
+-------------+--+
|     _c0     |
+-------------+--+
| helloworld  |
+-------------+--+
1 row selected (0.468 seconds)
```

整数类型和字符串类型的数据进行拼接。

```
jdbc:hive2://> select addition(name, age) from select_tbl;
+-----------+--+
|    _c0    |
+-----------+--+
| Sophie30  |
| Tom27     |
| Cocona38  |
| Sophie30  |
| Cocona38  |
| Tom27     |
+-----------+--+
6 rows selected (0.297 seconds)
```

（8）删除 UDF 的操作。

```
jdbc:hive2://> drop temporary function addition;
```

7.9 Hive 性能优化

为了高效地使用 Hive，需要对 Hive 进行性能优化。Hive 性能优化的常用方法如下。

（1）如果需要处理的数据量不大，可以使用 Hive 的本地模式，该模式比 Hadoop 的集群模式运行速度快。启用 Hive 本地模式的命令如下。

```
set hive.exec.mode.local.auto=true;                          //默认为 false
set hive.exec.mode.local.auto.inputbytes.max=50000000;       //输入字节数，默认为 128MB
set hive.exec.mode.local.auto.input.files.max=8;             //输入文件数，默认为 4
```

注意：为了便于讲解，命令后面增加了注解内容，使用命令时需要去掉注解。

（2）增加 Hive 的并行执行线程数，提高执行效率。

Hive 的查询包含一个或者多个阶段（Stage），这些阶段常用于执行 MapReduce 任务、抽样、合并、limit 等操作，Hive 一般一次执行一个阶段。当 Hive 执行多个阶段，并且阶段之间不存在依赖关系时，这些阶段可以并行执行，缩短执行时间。设置 Hive 的并行执行属性，并增加 Hive 并行执行线程数的命令如下。

```
set hive.exec.parallel=true;                    //默认为 false
set hive.exec.parallel.thread.number=16;        //默认可执行的最大并行线程数为 8
```

（3）在执行多表 Join 操作时，使小表在左，大表在右。原因是 Hive 在执行多表 Join 操作时，会先将数据缓存起来,然后和后面的表进行连接,如果小表在前,缓存数据相对较少。

(4)使用 Map Join。

如果一个表足够小,可以完全加载到内存中,那么 Hive 可以执行 Map Join。自动启用 Map Join 和设置小表大小的命令如下。

```
set hive.auto.convert.join=true;                    //自动启用 Map Join,默认为 false
set hive.mapjoin.smalltable.filesize=600000000;  //小表的大小默认为 25MB
```

(5)合理地设置桶表连接。

在大量数据进行 Join 时,使用 Map Join 会出现内存不足的情况,如果使用桶表连接(Bucket Map Join),就可以把少量桶的数据放到内存中进行 Map Join 操作。设置 Bucket Map Join 的命令如下。

```
set hive.auto.convert.join=true;              //自动启用 Map Join,默认为 false
set hive.optimize.bucketmapjoin = true        //默认为 false
```

(6)当能够使用半连接时,不要使用内连接。半连接一旦在右表中找到左表中的指定记录就立即停止扫描,效率更高。

(7)使用 limit 优化,返回执行整个语句后的部分结果。设置启用 limit 优化的命令如下。

```
set hive.limit.optimize.enable=true;
```

(8)如果数据分布不均匀,导致数据倾斜,可以采用以下优化方式。

```
set hive.optimize.skewjoin=true;     //默认为 false
set hive.skewjoin.key=100000;        //如果 Key 的个数超过该设定值,新的数据会发送给其他空闲的 Reduce
```

(9)group by 操作优化。

Hive 可以在 Map 阶段进行部分聚合操作,以此减少 Reduce 阶段的操作。而 group by 操作是一种可以在 Map 阶段进行的聚合操作。

设置在 Map 阶段进行聚合。

```
set hive.map.aggr=true;                              //默认为 true
```

设置 Map 阶段进行 group by 操作的条数。

```
set hive.groupby.mapaggr.checkinterval=100000;       //默认为 100000 条
```

Hive 在使用 group by 时经常发生数据倾斜,可以进行如下优化。

```
set hive.groupby.skewindata=true;                    //默认为 false
```

Hive 设置好以上属性后,在使用 group by 时,会触发一个额外的 MapReduce 作业,该作业 Map 阶段的输出数据将被随机地发送到 Reducer,以避免数据倾斜。

(10)合并小文件。

小于 HDFS 块大小的小文件数目过多,会大量占用 HDFS 的 NameNode 存储空间。合并小文件的命令如下。

```
set hive.merge.mapfiles=true                     //Map 阶段合并小文件
set hive.merge.mapredfiles=true                  //MapReduce 作业完成后合并小文件
set hive.merge.size.per.task=256000000           //设置作业完成后合并小文件的大小
set hive.merge.smallfiles.avgsize=16000000       //设置触发合并小文件的阈值
```

（11）设置 Reducer 个数。

Hive 在启用 MapReduce 作业时，Reducer 个数对作业执行效率的影响较大。如果未设置 Reducer 个数，Hive 会通过以下两个属性值来确定 Reducer 个数。

```
hive.exec.reducers.bytes.per.reducer    //①
hive.exec.reducers.max                  //②
```

Reducer 个数 N 的计算公式如下。

$$N=\min(②,总输入数据量/①)$$

在实际开发环境中，可以根据自身业务数据的特点设置上述两个属性值，具体命令示例如下。

```
jdbc:hive2://> hive.exec.reducers.bytes.per.reducer=300000000
jdbc:hive2://> hive.exec.reducers.max=16
```

（12）如果业务逻辑导致优化效果不明显，有时可以取出倾斜的数据进行处理，之后将处理完的数据与原数据进行合并。

7.10 Hive 案例分析

本节以分析机顶盒产生的用户收视数据为例，讲解如何使用 Hive。

（1）机顶盒产生的用户原始数据都有一定的格式，包含机顶盒号、收看的频道、收看的节目、收看的时间等信息。

（2）用户的原始数据通常不是直接交给 Hive 处理，而是先经过一个清洗和转化的过程。这个过程一般通过 Hadoop 作业来实现，目的是将数据转化成与 Hive 表对应的格式。

本案例的具体步骤如下。

1. 用户数据预处理

通过 MapReduce 作业将日志转化为固定的格式。

用户的原始数据如下所示。

```
< GHApp>< WIC   cardNum="1370695139"   stbNum="03111108020232488"   date="2012-09-21"
pageWidgetVersion="1.0">< A   e="13:55:11"   s="13:50:10"   n="104"   t="1"   pi="789"
p="%E5%86%8D%E5%9B%9E%E9%A6%96(21)" sn="BTV影视" />< /WIC>< /GHApp>
```

转化之后的数据如下所示，字段之间的分隔符为@。

```
1370695139@03111108020232488@2012-09-21@BTV影视@再回首@13:50:10@13:55:11@301
```

上面的字段分别代表：机顶盒号、用户编号、收看日期、频道、栏目、起始时间、结束时间、收视时长。

2. 创建 Hive 表

根据对应字段，使用 Hive 创建表。

```
create table tvdata(cardnum string,stbnum string,date string,sn string,p string ,s
string,e string,duration int) row format delimited fields terminated by '@' stored as textfile;
```

3. 将 HDFS 中的数据导入表

使用以下命令，将 HDFS 中的数据导入表。

```
load data inpath '/media/tvdata/part-r-00000' into table tvdata;
```

4. 编写 SQL 分析数据

使用 SQL 语句，统计每个频道的人均收视时长。

```
select sn,sum(duration)/count(*) from tvdata group by sn;
```

这里只是从一个角度分析数据，可以尝试从多个角度来分析数据。

7.11 本章小结

本章主要对 Hive 的原理、架构、安装步骤、数据类型、数据库和表的操作、函数进行了讲解。通过对 Hive 数据库和表的操作，读者可深刻体会 SQL 语句的易用性。熟悉 Hive 的相关知识有助于进一步加深对 MapReduce 和 HDFS 原理的理解，也为 Hive 与其他 Hadoop 组件的联合使用打下了基础。

7.12 习题

1. 填空题

（1）Hive 提供的操作简单的类 SQL 查询语言是_____。
（2）Hive 操作的真实数据存储在 Hadoop 文件系统中，通常是_____。
（3）切换到数据库 db02 的命令是_____。
（4）Hive 默认的文件格式是_____。
（5）_____关键字用来删除表的数据，保留表结构。

2. 选择题

（1）Hive 的大部分查询操作通过（　　）完成。

　　A．MapReduce 作业　　　　　　B．HDFS 作业
　　C．HBase 作业　　　　　　　　D．YARN 作业

（2）Hive 的复杂数据类型包括（　　）。

　　A．ARRAY　　B．MAP　　C．STRUCT　　D．UNION

（3）内部表和外部表的区别之一：创建表时是否使用（　　）关键字。

　　A．over　　B．outer　　C．exterior　　D．external

（4）（　　）关键字用来复制指定表的表结构和数据。

　　A．like　　B．copy　　C．as　　D．replication

（5）Hive 的用户自定义函数包括（　　）。

　　A．UDF　　B．UDAF　　C．UDTF　　D．UDHF

3. 思考题

（1）Hive 如何解决数据倾斜？
（2）Hive 的 Join 如何使用？

第8章 HBase分布式存储系统

本章学习目标
- 掌握 HBase 架构及其原理
- 掌握 HBase 的存储流程
- 熟悉 HBase 的安装和使用
- 理解 HBase 与 Hive 之间的关系

HBase 实现了在廉价硬件构成的集群中管理大规模数据。相对于关系型数据库，HBase 能够更灵活地通过增加节点的方式实现线程的扩展，更高效地处理分布式的大规模数据。当需要对大规模数据集进行实时读写、随机访问时，可以考虑使用 HBase。

8.1 认识 HBase

8.1.1 HBase 简介

HBase 是一个基于 Hadoop 的分布式、面向列的开源数据库，对大数据实现了随机定位和实时读写。

HBase 是基于 Google 的 Bigtable 技术实现的。Google Bigtable 利用 GFS 作为其文件存储系统，HBase 利用 Hadoop 的 HDFS 作为其文件存储系统；Google 运行 MapReduce 来处理 Bigtable 中的海量数据，HBase 同样利用 Hadoop 的 MapReduce 来处理 HBase 中的海量数据；Google Bigtable 利用 Chubby 进行协同服务，HBase 利用 ZooKeeper 进行协同服务。

HBase 具有以下特点。

（1）读取数据实时性强：可以实现对大数据的随机访问和实时读写。

（2）存储空间大：可以存储十亿行、百万列、上千个版本的数据。

（3）具有可伸缩性：可以通过增删节点实现数据的伸缩性存储。

（4）可靠性强：HBase 的 RegionServer 之间可以实现自动故障转移。

（5）面向列：面向列（族）的存储和权限控制，列（族）独立检索。

（6）数据类型单一：HBase 中的数据都是字符串，没有其他类型。

8.1.2 HBase 的数据模型

HBase 是一个面向列的数据库,数据模型主要有命名空间(Namespace)、表(Table)、行键(Rowkey)、列簇(Column Family)、列(Column)、时间戳(Timestamp)、单元格(Cell)。

1. 命名空间

命名空间可以对表进行逻辑分组,类似于关系型数据库系统中的数据库。

2. 表

表由行键和列簇组成,按行键的字典顺序进行排序。

3. 行键

行键是每一行数据的唯一标识,可以使用任意字符串表示。行键的最大长度为64KB,实际应用中一般为 10~1000 Byte。在 HBase 内部,行键保存为字节数组。

4. 列簇

列簇是列的集合,在创建表时必须声明列簇。一个列簇的所有列使用相同的前缀(列簇名称)。HBase 所谓的列式存储就是指数据按列簇进行存储,这种设计便于进行数据分析。

5. 列

列以键值对的形式进行存储。列的值是字节数组,没有类型和长度限定。列的格式通常为 column family:qualifier。例如,name:tom 列和 name:jack 列都是列簇 name 的成员,:后的内容通常称为限定符(Qualifier),限定符可以是任意的字节数组,相同列簇的限定符是唯一的,列的数量可以达到百万级别。

6. 单元格

单元格是由行键、列簇、版本唯一确定的单元。单元格中的数据全部以字节码形式存储。

7. 时间戳和版本

每个单元格通常保存着同一份数据的多个版本(Version),它们用时间戳来区分。时间戳是 64 位的整型数据。

时间戳可以被自动赋值和显式赋值。自动赋值是指在数据写入时,HBase 可以自动对时间戳进行赋值,该值是精确到毫秒的当前系统时间。显式赋值是指时间戳可以由客户显式指定。

为了方便地进行数据的版本管理,HBase 提供了两种数据版本回收方式。

(1)保存数据的最后 n 个版本。

(2)保存最近一段时间内的版本(例如,最近 7 天的版本)。

用户可以针对列簇进行自定义设置。

HBase 表的简单样式,如表 8.1 所示。

表 8.1　　HBase 表的简单样式

Rowkey	Column Family			Column Family			Column Family		
	col1	col2	col3	col1	col2	col3	col1	col2	col3
1									
2									
3									

8.1.3 HBase 架构

HBase 的架构如图 8.1 所示。

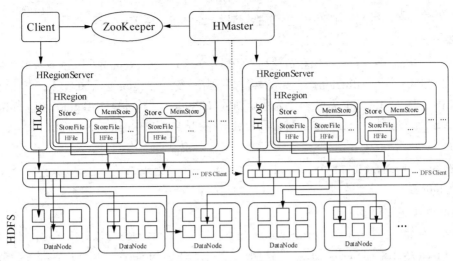

图 8.1　HBase 架构图

由 HBase 架构图可知，HBase 主要涉及 4 个模块：Client（客户端）、ZooKeeper、HMaster（主服务器）、HRegionServer（区域服务器）。其中，HRegionServer 模块包括 HRegion、Store、MemStore、StoreFile、HFile、HLog 等组件。下面对 HBase 涉及的模块和组件进行讲解。

1. Client

Client 通过 RPC 机制与 HBase 的 HMaster 和 HRegionServer 进行通信。Client 与 HMaster 进行管理类通信，与 HRegionServer 进行数据读写类通信。

2. ZooKeeper

ZooKeeper 在 HBase 中主要有以下两个方面的作用。

（1）HRegionServer 主动向 ZooKeeper 集群注册，使得 HMaster 可以随时感知各个 HRegionServer 的运行状态（是否在线），避免 HMaster 出现单点故障问题。

（2）HMaster 启动时会将 HBase 系统表加载到 ZooKeeper 集群，通过 ZooKeeper 集群可以获取当前系统表 hbase:meta 的存储所对应的 HRegionServer 信息。其中，系统表是指命名空间 hbase 下的表 namespace 和 meta。

3. HMaster

HMaster 负责维护表和 HRegion 的元数据信息，表的元数据信息保存在 ZooKeeper 上，HMaster 负载较小。HBase 一般有多个 HMaster，以便实现自动故障转移。HMaster 主要有以下几个方面的作用。

（1）管理用户对表的增、删、改、查操作。

（2）为 HRegionServer 分配 HRegion，负责 HRegionServer 的负载均衡。

（3）发现离线的 HRegionServer，并为其重新分配 HRegion。

（4）负责 HDFS 上的垃圾文件回收。

4. HRegionServer

HRegionServer 负责管理一系列 HRegion 对象，是 HBase 中的核心模块。一个 HRegionServer 一般会有多个 HRegion 和一个 HLog，用户可以根据实际需要添加或删除 HRegionServer。

HRegionServer 主要有以下两个方面的作用。

（1）维护 HMaster 分配的 HRegion，处理对这些 HRegion 的 I/O 请求。

（2）负责切分在运行过程中变得过大（默认超过 256MB）的 HRegion。

5. HRegion

HRegion 是 HBase 中分布式存储和负载均衡的最小单元。一个 HRegion 由一个或者多个 Store 组成。每个表起初只有一个 HRegion，随着表中数据不断增多，HRegion 会不断增大，增大到一定阈值（默认 256MB）时，HRegion 就会等分为两个新的 HRegion。不同的 HRegion 可以分布在不同的 HRegionServer 上，但同一个 HRegion 拆分后也会分布在相同的 HRegionServer 上。

6. Store

一个 Store 由 MemStore（1 个）和 StoreFile（0~n 个）组成。一个 Store 保存一个列簇。Store 是 HBase 存储的核心。

7. MemStore

MemStore 存储在内存中。当 MemStore 的大小达到一定阈值（默认 128MB）时，MemStore 会被刷新写入（Flush）磁盘文件，即生成一个快照。当关闭 HRegionServer 时，MemStore 会被强制刷新写入磁盘文件。

以下是 MemStore 刷新写入 StoreFile 文件的条件。

（1）达到 hbase.regionserver.global.MemStore.upperLimit，默认是 0.4，即堆内存的 40%。

（2）达到 hbase.hregion.MemStore.flush.size，默认是 128MB。

（3）达到 hbase.hregion.preclose.flush.size，默认是 5MB，并且确保 HRegion 已经关闭。

8. StoreFile

StoreFile 是 MemStore 中的数据写入磁盘后得到的文件。StoreFile 存储在 HDFS 上。

综上所述，HRegion 的组件之间的关系为：一个 HRegion 由一个或者多个 Store 组成；一个 Store 由 MemStore（1 个）和 StoreFile（0~n 个）组成；一个 Store 保存一个列簇；StoreFile 存储在 HDFS 上，MemStore 存储在内存中。

8.1.4 HBase 文件存储格式

HBase 的文件存储格式主要有 HFile 和 HLog 两种，下面分别进行介绍。

1. HFile

HFile 是 HBase 中键值数据的存储格式。HFile 文件是 Hadoop 的二进制格式的文件，StoreFile 底层存储使用的就是 HFile 文件。

一个 HFile 文件通常分解成多个块，对 HFile 文件的操作都是以块为单位进行的。HFile 文件的块大小可以在列簇级别中进行设置，推荐设置为 8KB~1024KB。较大的块有利于顺序读写数据，但由于需要解压更多的数据，不便于数据的随机读取。较小的块有利于随机读写，但需要占用更多内存，数据写入文件时相对较慢。

2. HLog

HLog 是 HBase 中 WAL（Write-Ahead-Log，预写日志）文件的存储格式。WAL 文件是 HBase 的 RegionServer 在处理数据过程中用来记录操作内容的一种日志文件。WAL 文件在物理上属于 Hadoop SequenceFile。

HBase 在处理数据时，先将数据临时保存在内存中，当数据量达到设置的阈值时，再将数据写入磁盘，这样做可以减少小文件。如果存储在内存中的数据由于设备故障等原因未及时写入磁盘，就会出现数据丢失的情况。WAL 文件解决了这种问题。如果设备出现故障，HBase 可以通过 WAL 文件恢复数据。

8.1.5 HBase 存储流程

下面分析 HBase 的整个存储流程。

（1）客户端提交变更操作（如插入、删除、计数新增）后，首先会连接上 ZooKeeper 找到 Root 表的存储位置，然后根据 meta 表的位置找到对应的 HRegion 所在的 HRegionServer。数据变更信息会先通过 HRegionServer 写入一个 commit log，也就是 WAL。写入 WAL 成功后，数据变更信息会存到 MemStore 中。MemStore 达到设定的阈值后，就会开始进行 Flush 操作，将其内容持久化到一个新的 HFile 中。在 Flush 操作过程中，MemStore 通过滚动机制继续对用户提供读写服务。随着 Flush 操作的不断进行，HFile 文件越来越多。HFile 文件超过设定的数量后，HBase 的内务处理机制就会将 HFile 小文件合并成更大的 HFile 文件。在合并的过程中，会进行版本的合并以及数据的删除。由于 StoreFile 是不变的，用户执行删除操作时，并不能简单地通过删除其键值对来删除数据内容。HBase 提供了删除标记机制，会告诉 HRegionServer 指定的键已经被删除了。这样其他用户检索这个键的内容时，也不会检索出来。在合并操作中，这些标记过的记录会被丢弃。经过多次合并，HFile 文件会越来越大，当达到设定的值时，会触发切片操作，将当前的 HRegion 根据行键对等切分成两个子 HRegion。当前的 HRegion 被废弃，两个子 HRegion 会被分配到其他 HRegionServer 上。通过 HMaster 的负载均衡调整，HRegion 会均匀分布到所有的 HRegionServer 中。

（2）当 HLog 满时，HRegionServer 就会启动 LogRoller 线程，通过执行 rollWriter()方法将所有序号小于最大序号的日志文件移动到.oldLog 目录中等待被删除。如果用户设置了延迟日志刷新，HRegionServer 会缓存有关表的所有变更，并通过 LogSyncer 类调用 sync()方法定时将变更信息同步到文件系统；否则，一旦有变更就会立刻同步到文件系统。

（3）一个 HRegionServer 中只有一个 WAL，所有 HRegion 共享此 WAL。HLog 会根据 HRegion 提交变更信息的先后顺序将操作依次写入 WAL。如果用户设置了 setWriteToWAL(false)方法，则有关表的所有 HRegion 变更日志都不会写入 WAL。

8.1.6 HBase 和 HDFS

1. HBase 和 HDFS 的联系

（1）HBase 管理的文件大部分存储在 HDFS 中。

（2）HBase 和 HDFS 都具有良好的容错性和扩展性，都可以扩展到成百上千个节点。

2. HBase 和 HDFS 的区别

（1）HDFS 是一个分布式的文件系统，适合对大规模数据进行批量处理，但不支持数据随机查找，不支持数据更新，不适合增量数据处理。

（2）HBase 是构建在 Hadoop 之上的大型数据库，支持数据的随机定位和实时读写，实现了处理数据的低延迟。

8.2 HBase 表设计

8.2.1 列簇设计

列簇设计的原则是：在合理范围内尽量减少列簇。

最优设计是将所有相关性很强的键值对都放在同一个列簇下，这样既能做到查询效率最高，又能保持尽可能少地访问不同的磁盘文件。

以用户信息为例，可以将基本信息存放在一个列簇，一些附加的额外信息存放在另一列簇。

8.2.2 行键设计

在 HBase 中，表会被划分为 1~n 个 HRegion，被托管在 RegionServer 中。HRegion 有两个重要的属性：StartKey 与 EndKey，表示这个 HRegion 维护的行键范围。当我们要读/写数据时，如果行键落在某个范围内，就能定位到目标 HRegion 并读/写相关的数据。能否快速精准地定位到我们想要处理的数据，取决于行键的设计。

行键设计有三大原则，分别为长度原则、散列原则和行键唯一原则。

1. 长度原则

行键是一个二进制码，很多开发者建议把行键的长度设计为 10~100 个字节，其实是越短越好，最好不要超过 16 个字节。原因如下。

（1）数据的持久化文件 HFile 按照键值存储数据，如果行键过长，比如 100 个字节，1000 万列数据仅行键就要占用 100×1000 万=10 亿个字节，将近 1GB 数据，这会极大影响 HFile 的存储效率。

（2）MemStore 将缓存部分数据到内存，如果行键过长，内存的有效利用率会降低，系统将无法缓存更多的数据，这会降低检索效率。因此行键的字节长度越短越好。

2. 散列原则

加盐（Salting）：在行键的前面增加随机数，也就是给行键分配一个随机前缀，使行键的开头不同。分配的前缀种类数量应该和你想使用的 HRegion 的数量一致。散列之后的行键会根据前缀分散到各个 HRegion 上，以避免热点。

哈希（Hashing）：哈希会使同一行永远用一个前缀散列。使用确定的哈希让客户端可以重构完整的行键，可以使用 get 操作准确获取某一个行数据。

行键反转（RowKey Reverse）：反转固定长度或者数字格式的行键，把行键中经常改变的部分（最没有意义的部分）放在前面。以手机号为例，可以将手机号反转后的字符串作为行键，这样的就避免了固定开头导致热点问题。

3. 行键唯一原则

必须在设计上保证行键的唯一性。行键是按照字典顺序排序存储的，设计行键的时候，要充分利用这个排序的特点，将经常读取的数据存储到同一块，将最近可能会访问的数据存储到同一块。

8.3 HBase 安装

本书使用的是 1.2.6 版本的 HBase，安装包见附录。HBase 的安装模式有单机模式、伪分布式、完全分布式、HA 模式。本书主要讲述 HBase 的单机模式、HA 模式的安装步骤。

8.3.1 HBase 的单机模式

HBase 的单机模式安装步骤如下。

（1）将 HBase 安装包 hbase-1.2.6-bin.tar.gz 放到虚拟机 qf01 的/root/Downloads/目录下，切换到 root 用户，解压 HBase 安装包到/mysoft 目录下。

```
[root@qf01 ~]# tar -zxvf /root/Downloads/hbase-1.2.6-bin.tar.gz -C /mysoft/
```

（2）切换到/mysoft目录下，将hbase-1.2.6重命名为hbase。

```
[root@qf01 ~]# cd /mysoft/
[root@qf01 mysoft]# mv hbase-1.2.6 hbase
```

（3）打开/etc/profile文件，配置HBase环境变量。

```
[root@qf01 mysoft]# vi /etc/profile
```

在文件末尾添加如下三行内容。

```
# HBase environment variables
export HBASE_HOME=/mysoft/hbase
export PATH=$PATH:$HBASE_HOME/bin
```

（4）使环境变量生效。

```
[root@qf01 mysoft]# source /etc/profile
```

（5）切换到/soft/hbase/conf目录下，修改文件hbase-env.sh。

```
[root@qf01 mysoft]# cd /mysoft/hbase/conf
[root@qf01 conf]# vi hbase-env.sh
```

（6）将# export JAVA_HOME=/usr/java/jdk1.6.0/一行替换为如下内容。

```
export JAVA_HOME=/usr/local/jdk1.8.0_121
```

（7）修改hbase-site.xml文件，将<configuration>和</configuration>两行替换为如下内容。

```
<configuration>
    <!-- 指定HBase存放数据的目录 -->
    <property>
        <name>hbase.rootdir</name>
        <value>file:///root/hbasedir/hbase</value>
    </property>
        <!-- 指定ZooKeeper集群存放数据的目录 -->
    <property>
        <name>hbase.zookeeper.property.dataDir</name>
        <value>/root/hbasedir/hbase/zkdir</value>
    </property>
</configuration>
```

（8）启动HBase单机模式。

```
[root@qf01 conf]# start-hbase.sh
```

（9）使用jps命令查看HBase进程。

```
[root@qf01 conf]# jps
8260 Jps
8157 HMaster
```

HMaster进程就是HBase的主进程。HMaster进程启动表明HBase单机模式启动成功。
（10）查看HBase的Web界面（如图8.2所示），在浏览器中输入http://192.168.142.131:16010。
（11）关闭HBase。

```
[root@qf01 conf]# stop-hbase.sh
```

```
stopping hbase..................
```

关闭 HBase 时会出现自动打点,需要耐心等待。

图 8.2 HBase 的 Web 界面

8.3.2 HBase 的 HA 模式

下面从安装规划、安装步骤、启动和关闭 HBase 进程三个方面,介绍 HBase 的 HA 模式。

1. 安装规划

HBase 的 HA 模式安装规划,如表 8.2 所示。

表 8.2　　　　　　　　　　　　HBase 的 HA 模式安装规划

主机名	IP 地址	相关进程
qf01	192.168.142.131	NameNode、DataNode、NodeManager、ResourceManager DFSZKFailoverController、QuorumPeerMain、JournalNode、HMaster、HRegionServer
qf02	192.168.142.132	NameNode、DataNode、NodeManager、ResourceManager DFSZKFailoverController、QuorumPeerMain、JournalNode、HMaster、HRegionServer
qf03	192.168.142.133	DataNode、NodeManager、QuorumPeerMain、JournalNode、HRegionServer

2. 安装步骤

HBase 的 HA 模式安装步骤如下。

(1)在虚拟机 qf01 上修改/mysoft/hbase/conf/hbase-site.xml 文件,将<configuration>至</configuration>的内容替换如下。

```
<configuration>
    <!-- 开启 HBase 的完全分布式 -->
    <property>
        <name>hbase.cluster.distributed</name>
        <value>true</value>
    <!-- 指定 ZooKeeper 集群存放数据的目录 -->
    <property>
        <name>hbase.zookeeper.property.dataDir</name>
        <value>/home/hadoopdata/hbase/zkdir</value>
    </property>
    <!-- 指定 HBase 需要连接的 zookeeper 集群 -->
    <property>
        <name>hbase.zookeeper.quorum</name>
        <value>qf01,qf02,qf03</value>
    </property>
```

```
</configuration>
```

（2）修改/mysoft/hbase/conf/hbase-env.sh 文件。

① 将# export HBASE_MANAGES_ZK=true 一行替换为如下内容，目的是设置 HBase 不使用内置的 ZooKeeper，而使用外部安装的 ZooKeeper 集群。

```
export HBASE_MANAGES_ZK=false
```

② 将如下内容删除，或者在两个 export 前各添加一个#（改为注释）。

```
export HBASE_MASTER_OPTS="$HBASE_MASTER_OPTS -XX:PermSize=128m -XX:MaxPermSize=128m"
export HBASE_REGIONSERVER_OPTS="$HBASE_REGIONSERVER_OPTS -XX:PermSize=128m -XX:MaxPermSize=128m"
```

（3）修改/mysoft/hbase/conf/regionservers 文件，将文件中的内容替换如下。

```
qf01
qf02
qf03
```

（4）在/mysoft/hbase/conf 目录下新建文件 backup-masters，用于备份 HBase 的主节点 qf01，当主节点崩溃时，HBase 自动启用备份节点。

```
[root@qf01 conf]# vi backup-masters
```

在 backup-masters 中添加如下内容。

```
qf02
```

（5）将 Hadoop 的配置文件目录（/usr/local/hadoop-2.7.3/etc/hadoop）下的 core-site.xml 和 hdfs-site.xml 复制到/mysoft/hbase/conf 目录下。

```
[root@qf01 conf]# cd /usr/local/hadoop-2.7.3/etc/hadoop/
[root@qf01 hadoop]# cp core-site.xml hdfs-site.xml /mysoft/hbase/conf
```

（6）将 HBase 的安装目录复制到虚拟机 qf02 和 qf03。

```
[root@qf01 hadoop]# scp -r /mysoft/hbase qf02:/mysoft/hbase
[root@qf01 hadoop]# scp -r /mysoft/hbase qf03:/mysoft/hbase
```

（7）启动 HBase 的 HA 模式，需要先启动 ZooKeeper 和 Hadoop 集群。

```
[root@qf01 ~]# xzk.sh start
[root@qf01 ~]# start-dfs.sh
[root@qf01 ~]# start-yarn.sh
[root@qf01 ~]# start-hbase.sh
starting master, logging to /mysoft/hbase/logs/hbase-root-master-qf01.out
  qf03: starting regionserver, logging to /mysoft/hbase/bin/../logs/hbase-root-regionserver-qf03.out
  qf02: starting regionserver, logging to /mysoft/hbase/bin/../logs/hbase-root-regionserver-qf02.out
  qf01: starting regionserver, logging to /mysoft/hbase/bin/../logs/hbase-root-regionserver-qf01.out
  qf02: starting master, logging to /mysoft/hbase/bin/../logs/hbase-root-master-qf02.out
```

（8）使用 jps 命令查看进程。

```
[root@qf01 conf]# xcmd.sh jps
============ qf01 jps ============
```

```
5570 ResourceManager
2339 QuorumPeerMain
4887 NameNode
6183 HMaster
5224 JournalNode
5417 DFSZKFailoverController
5705 NodeManager
6329 HRegionServer
5004 DataNode
6623 Jps
============ qf02 jps ============
3984 DFSZKFailoverController
4082 ResourceManager
4709 Jps
4167 NodeManager
3658 NameNode
3851 JournalNode
3741 DataNode
4382 HRegionServer
4462 HMaster
2367 QuorumPeerMain
============ qf03 jps ============
3664 NodeManager
3425 DataNode
4034 Jps
3860 HRegionServer
2541 QuorumPeerMain
3535 JournalNode
```

出现以上进程表明 HBase 的 HA 模式启动成功。

3. 启动和关闭 HBase 进程

启动和关闭 HBase 进程的常用命令及含义，如表 8.3 所示。

表 8.3　　　　　　　启动和关闭 Hadoop 进程的常用命令及含义

命令	含义
start-hbase.sh	启动 HBase 的所有 HMaster、HRegionserver、备份的 HMaster 进程
stop-hbase.sh	关闭 HBase 的所有 HMaster、HRegionserver、备份的 HMaster 进程
hbase-daemon.sh start master	单独启动 HMaster 进程
hbase-daemon.sh stop master	单独关闭 HMaster 进程
hbase-daemons.sh start regionserver	启动所有的 HRegionserver 进程
hbase-daemons.sh stop regionserver	关闭所有的 HRegionserver 进程
hbase-daemons.sh start master-backup	启动所有备份的 HMaster 进程
hbase-daemons.sh stop master-backup	关闭所有备份的 HMaster 进程

8.4　HBase Shell 常用操作

　　HBase 提供了一个与用户交互的 Shell 终端，HBase 的一部分运维工作需要通过 Shell 终端来完成。HBase Shell 常用的操作命令主要有常规命令（General）、命名空间相关命令、数据定义语言（Data Definition Language，DDL）命令和数据操作语言（Data Manipulation Language，DML）命令。下面介绍这四类命令的相关操作。

1. 常用基本操作

（1）启动 HBase Shell 命令行。

```
[root@qf01 ~]# hbase shell
..
hbase(main):001:0>
```

hbase(main):001:0>中的 001 用来统计用户输入的行数，该数字会随着用户输入的行数自增，为了方便讲述，以下统一使用如下内容。

```
hbase(main)>
```

（2）关闭 HBase Shell 命令行。

关闭 HBase Shell 命令行，可以使用以下任一命令。

```
hbase(main)> quit
hbase(main)> exit
```

（3）查看服务器状态。

```
hbase(main)> status
1 active master, 1 backup masters, 3 servers, 0 dead, 1.6667 average load
```

由返回结果可知，目前有 1 个活跃的主节点，1 个备份的节点，3 个服务节点。

（4）查看 HBase 的版本。

```
hbase(main)> version
1.2.6, rUnknown, Mon May 29 02:25:32 CDT 2017
```

（5）查看当前使用 HBase 的用户。

```
hbase(main)> whoami
root (auth:SIMPLE)
    groups: root
```

（6）查看 HBase Shell 的帮助信息。

```
hbase(main)> help
```

在返回的帮助信息中，列出了 HBase Shell 的所有命令。通过 help 'command'命令可以查看 command 的详细用法。

查看 status 命令的用法，示例如下。

```
hbase(main)> help 'status'
Show cluster status. Can be 'summary', 'simple', 'detailed', or 'replication'. The
default is 'summary'. Examples:
  hbase> status
  hbase> status 'simple'
  ...
```

2. 常用的命名空间操作

（1）创建命名空间 ns（ns 为命名空间的名称）。

```
hbase(main)> create_namespace 'ns'
```

（2）查看所有的命名空间。

```
hbase(main)> list_namespace
```

```
NAMESPACE
default
hbase
ns
3 row(s) in 0.0470 seconds
```

由返回结果可知,HBase 默认定义两个命名空间:hbase 和 default。hbase 包括两个系统表 namespace 和 meta。default 主要存放用户建表时未指定命名空间的表。

(3)查看指定命名空间下的所有表。

```
hbase(main)> list_namespace_tables 'hbase'
TABLE
meta
namespace
2 row(s) in 0.0310 seconds
```

(4)删除命名空间 ns。

```
hbase(main)> drop_namespace 'ns'
```

3. 常用的 DDL 操作

DDL 主要用于与管理表相关的操作,常用的 DDL 操作有创建表、修改表、启用和禁用表、删除表、罗列表等。

(1)创建表。

创建表时需要指定表名和列簇。

① 在自定义的命名空间下创建表,语法如下。

```
create 'namespace名称:表名称','列簇名称1','列簇名称2'...
```

在命名空间 ns 下创建表 t1,并添加 3 个列簇 f1、f2、f3。

```
hbase(main)> create_namespace 'ns'
hbase(main)> create 'ns:t1', 'f1', 'f2', 'f3'
0 row(s) in 1.4040 seconds
=> Hbase::Table - ns:t1
```

② 在 HBase 自带的命名空间 default 下创建表,语法如下。

```
create '表名称','列簇名称1','列簇名称2'...
```

创建表 t2,并添加 2 个列簇 f1、f2。

```
hbase(main)> create 't2', 'f1', 'f2'
```

(2)查看所有的表。

```
hbase(main)> list
TABLE
ns:t1
t2
2 row(s) in 0.0100 seconds
=> ["ns:t1", "t2"]
```

(3)查看指定表是否存在。

```
hbase(main)> exists 't2'
Table t2 does exist
```

```
0 row(s) in 0.0190 seconds
```

（4）查看表描述。

查看表描述主要是指查看表是否可用和表的元数据信息。查看表描述的关键字是 describe，可以简写为 desc。

```
hbase(main)> desc 't2'
Table t2 is ENABLED
t2
COLUMN FAMILIES DESCRIPTION
 {NAME => 'f1', BLOOMFILTER => 'ROW', VERSIONS => '1', IN_MEMORY => 'false',
KEEP_DELETED_CELLS => 'FALSE', D
 ATA_BLOCK_ENCODING => 'NONE', TTL => 'FOREVER', COMPRESSION => 'NONE', MIN_VERSIONS =>
'0', BLOCKCACHE => 't
 rue', BLOCKSIZE => '65536', REPLICATION_SCOPE => '0'}
 {NAME => 'f2', BLOOMFILTER => 'ROW', VERSIONS => '1', IN_MEMORY => 'false',
KEEP_DELETED_CELLS => 'FALSE', D
 ATA_BLOCK_ENCODING => 'NONE', TTL => 'FOREVER', COMPRESSION => 'NONE', MIN_VERSIONS =>
'0', BLOCKCACHE => 't
 rue', BLOCKSIZE => '65536', REPLICATION_SCOPE => '0'}
2 row(s) in 0.0560 seconds
```

其中，{}中的内容是列簇的描述信息，简介如下。

```
NAME => 'f1'                          //列簇的名称为 f1
BLOOMFILTER => 'ROW'                  //布隆过滤器的类型为 ROW
VERSIONS => '1'                       //保存的版本数为 1
IN_MEMORY => 'false'                  // IN_MEMORY 常驻 cache
KEEP_DELETED_CELLS => 'FALSE'         //是否保留被删除的 Cell
DATA_BLOCK_ENCODING => 'NONE'         //数据块编码为 NONE，表示不使用数据块编码
TTL => 'FOREVER'                      //存在时间值，HBase 将在到期时删除行
COMPRESSION => 'NONE'                 //压缩算法为 NONE，表示不使用压缩
MIN_VERSIONS => '0'                   //最小版本的默认值为 0，表示该功能已被禁用
BLOCKCACHE => 'true'                  //数据块缓存属性
BLOCKSIZE => '65536'                  //HFile 数据块大小，默认为 64KB
REPLICATION_SCOPE => '0'              //是否复制列簇，REPLICATION_SCOPE 的值为 0 或 1，0 表示禁
用复制，1 表示启用复制
```

（5）启用表。

① 新建表完成后，表默认处于启用状态。

```
hbase(main)> enable 't2'
```

② 判断指定表是否被启用。

```
hbase(main)> is_enabled 't2'
true
```

（6）禁用表。

① 在删除表之前，需要禁用表。

```
hbase(main)> disable 'ns:t1'
```

② 判断指定表是否被禁用。

```
hbase(main)> is_disabled 't2'
```

```
false
```

（7）向表中添加列簇。

```
hbase(main)> alter 't2', 'f3'
hbase(main)> desc 't2'
```

（8）删除表。

```
hbase(main)> drop 'ns:t1'
```

注意：删除表之前需要先禁用表，ns:t1 已经在（6）中被禁用。

4. 常用的 DML 操作

DML 主要用于与处理数据相关的操作，常用的 DML 操作有数据的添加、删除、修改、查询等。

（1）向表中添加或更新数据，语法如下。

```
put '表名称', 'Rowkey 名称', '列簇名称:列名称', '值'
```

① 向表中添加数据。

```
hbase(main)> create 't1', 'f1', 'f2', 'f3'
hbase(main)> put 't1','row1','f1:id','1'
hbase(main)> put 't1','row1','f1:name','tom'
hbase(main)> put 't1','row1','f1:age','21'
hbase(main)> put 't1','row1','f2:id','2'
hbase(main)> put 't1','row1','f2:name','jack'
hbase(main)> put 't1','row1','f2:age','22'
hbase(main)> put 't1','row2','f1:city','Shanghai'
hbase(main)> put 't1','row3','f1:country','China'
```

② 更新 f1:city 列的数据。

```
hbase(main)> put 't1','row2','f1:city','Beijing'
```

（2）查看表中的数据，语法如下。

```
scan '表名称',{COLUMNS => ['列簇名称:列名称', ...], LIMIT => 行数}
```

① 查看指定表的所有数据。

```
hbase(main)> scan 't1'
ROW                COLUMN+CELL
 row1              column=f1:age, timestamp=1537794177479, value=21
 row1              column=f1:id, timestamp=1537794159437, value=1
 row1              column=f1:name, timestamp=1537794173185, value=tom
 row1              column=f2:age, timestamp=1537794191053, value=22
 row1              column=f2:id, timestamp=1537794181776, value=2
 row1              column=f2:name, timestamp=1537794186445, value=jack
 row2              column=f1:city, timestamp=1537794791005, value=Beijing
 row3              column=f1:country, timestamp=1537794798036, value=China
3 row(s) in 0.0170 seconds
```

② 查看指定表的指定列的所有数据。

```
hbase(main)> scan 't1', {COLUMNS => ['f1:id', 'f2:id']}
ROW                COLUMN+CELL
 row1              column=f1:id, timestamp=1537794159437, value=1
 row1              column=f2:id, timestamp=1537794181776, value=2
1 row(s) in 0.0410 seconds
```

③ 查看指定表的指定列的前 *n* 行数据。

```
hbase(main)> scan 't1', {COLUMNS => ['f1:id', 'f1:name', 'f1:age', 'f1:city',
'f1:country'], LIMIT => 2}
ROW                  COLUMN+CELL
row1                 column=f1:age, timestamp=1537794177479, value=21
row1                 column=f1:id, timestamp=1537794159437, value=1
row1                 column=f1:name, timestamp=1537794173185, value=tom
row2                 column=f1:city, timestamp=1537795506779, value=Beijing
2 row(s) in 0.0270 seconds
```

（3）获取表中指定行键下的数据。

① 获取指定表的指定行键下的所有数据，语法如下。

```
get '表名称', 'Rowkey'
```

示例如下。

```
hbase(main)> get 't1', 'row2'
COLUMN                   CELL
f1:city                  timestamp=1537795506779, value=Beijing
1 row(s) in 0.0170 seconds
```

② 获取指定表的指定行键的指定列簇下所有数据，语法如下。

```
get '表名称', 'Rowkey', '列簇名称'
```

示例如下。

```
hbase(main)> get 't1', 'row1', 'f2'
```

③ 获取指定表的指定行键的指定列的数据，语法如下。

```
get '表名称', 'Rowkey', '列名称'
```

示例如下。

```
hbase(main)> get 't1', 'row1', 'f2:name'
```

（4）统计表的总行数。

```
hbase(main)> count 't1'
```

（5）删除数据。

① 删除一个单元格的数据，语法如下。

```
delete '表名称', 'Rowkey', '列名称', 时间戳（timestamp）
```

示例如下。

```
hbase(main)> delete 't1', 'row1', 'f1:age', 1537794177479
```

② 删除指定列的数据，语法如下。

```
delete '表名称', 'Rowkey', '列名称'
```

示例如下。

```
hbase(main)> delete 't1', 'row1', 'f1:name'
```

③ 删除指定行的所有数据，语法如下。

```
hbase(main)> deleteall 't1', 'row3'
```

④ 删除表中所有数据，使用关键字 **truncate**，示例如下。

```
hbase(main)> put 't2','row1','f1:id','8'
hbase(main)> put 't2','row1','f1:name','sophie'
hbase(main)> truncate 't2'
Truncating 't2' table (it may take a while):
 - Disabling table...
 - Truncating table...
0 row(s) in 3.7270 seconds
hbase(main)> scan 't2'
ROW                 COLUMN+CELL
0 row(s) in 0.0340 seconds
```

由返回信息可知，HBase 先将表禁用，再删除表中所有数据。

8.5 HBase 编程

HBase 是用 Java 语言编写的，下面讲解通过 Java API 来操作 HBase。

8.5.1 配置开发环境

（1）在 IntelliJ IDEA 中项目 testHadoop 下新建模块 testHBase，修改 pom.xml 文件内容如下。

```xml
<?xml version="1.0" encoding="UTF-8"?>
<project xmlns="http://maven.apache.org/POM/4.0.0"
         xmlns:xsi="http://www.w3.org/2001/XMLSchema-instance"
         xsi:schemaLocation="http://maven.apache.org/POM/4.0.0 http://maven.apache.org/xsd/maven-4.0.0.xsd">
    <modelVersion>4.0.0</modelVersion>
    <groupId>com.qf</groupId>
    <artifactId>testHBase</artifactId>
    <version>1.0-SNAPSHOT</version>
    <dependencies>
        <dependency>
            <groupId>org.apache.hbase</groupId>
            <artifactId>hbase-client</artifactId>
            <version>1.2.6</version>
        </dependency>
        <dependency>
            <groupId>org.apache.hbase</groupId>
            <artifactId>hbase-server</artifactId>
            <version>1.2.6</version>
        </dependency>
        <dependency>
            <groupId>junit</groupId>
            <artifactId>junit</artifactId>
            <version>4.12</version>
        </dependency>
    </dependencies>
</project>
```

（2）将虚拟机 qf01 的 /mysoft/hbase/conf 目录下的 hbase-site.xml、core-site.xml、hdfs-site.xml 三

个文件复制到 IntelliJ IDEA 中 testHBase 模块的 src/main/resources 目录下。

8.5.2 使用 Java API 操作 HBase

【例 8-1】HBaseTest.java

```java
1    //使用 Java API 操作 HBase
2    public class HBaseTest {
3        // 1.创建 Namespace    2.查看所有的 Namespace
4        @Test
5        public void testCreateNamespace() throws IOException {
6            //获取 HBase 的配置信息
7            Configuration conf = HBaseConfiguration.create();
8            conf.set("fs.defaultFS", "hdfs://192.168.142.131:8020");
9            conf.set("hbase.zookeeper.quorum", "qf01,qf02,qf03");
10           //创建连接
11           Connection conn = ConnectionFactory.createConnection(conf);
12           //得到对 Namespace 和表进行操作的管理员权限
13           Admin admin = conn.getAdmin();
14           //创建 NamespaceDescriptor 实例,并设置 Namespace 名称为 ns1
15           NamespaceDescriptor nsD = NamespaceDescriptor.create("ns1").build();
16           //创建 Namespace
17           admin. testCreateNamespace(nsD);
18           //查看所有 Namespace
19           NamespaceDescriptor[] nsDs = admin.listNamespaceDescriptors();
20           for (NamespaceDescriptor nsd : nsDs) {
21               System.out.println(nsd);
22           }
23           admin.close();
24       }
```

运行结果如下。

```
{NAME => 'default'}
{NAME => 'hbase'}
{NAME => 'ns'}
{NAME => 'ns1'}
```

判断表是否存在。

```java
1    /**
2     * 1.判断表是否存在
3     * 2.创建表
4     * 3.查看所有的表
5     */
6    @Test
7    public void operateTable() throws Exception {
8        Configuration conf = HBaseConfiguration.create();
9        conf.set("fs.defaultFS", "hdfs://192.168.142.131:8020");
10       conf.set("hbase.zookeeper.quorum", "qf01,qf02,qf03");
11       Connection conn = ConnectionFactory.createConnection(conf);
12       Admin admin = conn.getAdmin();
13       //设置表名称
14       TableName tn = TableName.valueOf("ns1:t1");
```

```
15        //判断表是否存在,如果存在就删除
16        if (admin.tableExists(tn)) {
17            if (admin.isTableEnabled(tn)) {
18                admin.disableTable(tn);
19            }
20            admin.deleteTable(tn);
21        }
22        //创建 HTableDescriptor 对象,并添加表名称
23        HTableDescriptor table = new HTableDescriptor(tn);
24        //创建 HColumnDescriptor 对象,并添加列簇名称
25        HColumnDescriptor cf1 = new HColumnDescriptor("cf1");
26        HColumnDescriptor cf2 = new HColumnDescriptor("cf2");
27        //在表中添加列簇
28        table.addFamily(cf1);
29        table.addFamily(cf2);
30        //创建表
31        admin.createTable(table);
32        //查看所有的表
33        TableName[] tns = admin.listTableNames();
34        for (TableName tableName : tns) {
35            System.out.println(tableName);
36        }
37        admin.close();
38    }
```

运行结果如下。

```
ns1:t1
t1
t2
```

添加数据到表中。

```
1    /**
2     * 1.添加数据到表中
3     * 2.查看指定表的所有数据
4     */
5    @Test
6    public void putData() throws Exception {
7        Configuration conf = HBaseConfiguration.create();
8        conf.set("fs.defaultFS", "hdfs://192.168.142.131:8020");
9        conf.set("hbase.zookeeper.quorum", "qf01,qf02,qf03");
10       Connection conn = ConnectionFactory.createConnection(conf);
11       //获取表对象
12       Table table = conn.getTable(TableName.valueOf("ns1:t1"));
13       //创建 put 实例
14       Put data = new Put(Bytes.toBytes("row1"));
15       //添加列数据
16       data.addColumn(Bytes.toBytes("cf1"), Bytes.toBytes("age"), Bytes.toBytes("18"));
17       //添加数据到表中
18       table.put(data);
19       //查看指定表的所有数据
```

```
20            //创建 scan 对象
21            Scan scan = new Scan();
22            //通过扫描器得到结果集
23            ResultScanner rs = table.getScanner(scan);
24            //得到迭代器
25            Iterator<Result> it = rs.iterator();
26            printData(it);
27            table.close();
28            conn.close();
29        }
```

运行结果如下。

```
row1,cf1:age,18
```

列过滤器编程。

```
1     //迭代输出每行的所有数据
2     public static void printData(Iterator<Result> it) {
3         while (it.hasNext()) {
4             Result next = it.next();
5             List<Cell> cells = next.listCells();
6             for (Cell cell : cells) {
7                 String row = Bytes.toString(CellUtil.cloneRow(cell));
8                 String cf = Bytes.toString(CellUtil.cloneFamily(cell));
9                 String qualifier = Bytes.toString(CellUtil.cloneQualifier(cell));
10                String value = Bytes.toString(CellUtil.cloneValue(cell));
11                System.out.println(row + "," + cf + ":" + qualifier + "," + value);
12            }
13        }
14    }
15 }
```

分别运行例 8-1 中的 testCreateNamespace()、operateTable()、putData()三个方法，即可验证运行结果。除了以上示例中涉及的操作，读者还可以尝试使用 API 编程实现 HBase Shell 中的其他常用操作。

8.5.3 使用 HBase 实现 WordCount

下面使用 HBase 实现 WordCount。

（1）创建 HBase 表 wordcount，具体命令如下。

```
hbase(main)> create 'wordcount', 'f1'
```

（2）向 wordcount 表中插入数据。

```
hbase(main)> put 'wordcount', 'row1', 'f1:word', 'hello xiao qian'
hbase(main)> put 'wordcount', 'row2', 'f1:word', 'hello hadoop'
hbase(main)> put 'wordcount', 'row3', 'f1:word', 'hello hbase'
```

（3）实现 Mapper。

【例 8-2】MyMapper.java

```
1  public class MyMapper extends Mapper<ImmutableBytesWritable, Result, Text, IntWritable> {
2      @Override
3      protected void map(ImmutableBytesWritable key, Result value, Context context)
```

```
throws IOException, InterruptedException {
    4            //1.获得结果集中的Cell集合
    5            List<Cell> cells = value.listCells();
    6            //2.迭代Cell集合
    7            for (Cell cell : cells) {
    8                //取出Cell中的值
    9                String words = Bytes.toString(CellUtil.cloneValue(cell));
   10                //将多个单词拆分放入数组
   11                String[] arr = words.split(" ");
   12                //迭代数组，取出单个单词
   13                for (String word : arr) {
   14                    //将输出的K-V对存入context
   15                    context.write(new Text(word), new IntWritable(1));
   16                }
   17            }
   18        }
   19    }
```

（4）实现 Reducer。

【例 8-3】MyReducer.java

```
    1    public class MyReducer extends Reducer<Text, IntWritable, Text, IntWritable> {
    2        @Override
    3        protected void reduce(Text key, Iterable<IntWritable> values, Context context)
throws IOException, InterruptedException {
    4            //1.定义一个计数器
    5            int count = 0;
    6            //2.迭代数组，将输出的K-V对存入context
    7            for (IntWritable i : values) {
    8                count = count + i.get();
    9            }
   10            context.write(key, new IntWritable(count));
   11        }
   12    }
```

（5）创建 MapReduce 作业。

【例 8-4】HBaseWordCountApp.java

```
    1    public class HBaseWordCountApp {
    2        public static void main(String[] args) throws Exception {
    3            //1.新建配置对象，为配置对象设置文件系统
    4            Configuration conf = new Configuration();
    5            conf.set("fs.defaultFS", "hdfs://192.168.142.131:8020");
    6            conf.set(TableInputFormat.INPUT_TABLE, "wordcount");
    7            //2.添加ZooKeeper客户端主机地址
    8            conf.set("hbase.zookeeper.quorum", "qf01,qf02,qf03");
    9            //3.设置Job属性
   10            Job job = Job.getInstance(conf, "HBaseWordCount");
   11            job.setJarByClass(HBaseWordCountApp.class);
   12            //4.设置数据输入路径
   13            Path inPath = new Path("/tmp/hbase-root/hbase/data/default/wordcount");
   14            FileInputFormat.addInputPath(job, inPath);
   15            //5.设置输入格式
```

```
16              job.setInputFormatClass(TableInputFormat.class);
17              //6.设置 Job 执行的 Mapper 类
18              job.setMapperClass(MyMapper.class);
19              //7.设置 Job 执行的 Reducer 类和输出 K-V 类型
20              job.setReducerClass(MyReducer.class);
21              job.setOutputKeyClass(Text.class);
22              job.setOutputValueClass(IntWritable.class);
23              //8.设置数据输出路径
24              Path outPath = new Path("/outdata/hbasewordcount");
25              FileOutputFormat.setOutputPath(job, outPath);
26              //9.MapReduce 作业完成后退出系统
27              System.exit(job.waitForCompletion(true) ? 0 : 1);
28          }
29      }
```

(6)运行 MapReduce 作业。在虚拟机 qf01 上查看 HDFS 文件/outdata/hbasewordcount/part-r-00000,具体命令如下。

```
[root@qf01 ~]# hdfs dfs -cat /outdata/hbasewordcount/part-r-00000
hadoop   1
hbase    1
hello    3
qian     1
xiao     1
```

出现上述结果,表明使用 HBase 成功实现了 WordCount。

8.6 HBase 过滤器和比较器

8.6.1 过滤器

过滤器负责在服务端判断数据是否满足指定条件,并将满足条件的数据返回给客户端,类似于 SQL 语句中的 WHERE 子句。过滤器的作用是减少服务器通过网络返回到客户端的数据量。

1. 过滤器种类

过滤器种类较多,本书重点讲述两类过滤器:比较过滤器和专用过滤器。

(1)比较过滤器(Compared Filter)有行过滤器(RowFilter)、列簇过滤器(FamilyFilter)、列过滤器(QualifierFilter)、值过滤器(ValueFilter)。

(2)专用过滤器(Dedicated Filter)有单列值过滤器(SingleColumnValueFilter)、单列值排除过滤器(SingleColumnValueExcludeFilter)、前缀过滤器(PrefixFilter)、列前缀过滤器(ColumnPrefixFilter)、分页过滤器(PageFilter)。

- 单列值过滤器通过搜索指定列和列值,返回含有指定列和列值的整行数据。
- 单列值排除过滤器通过搜索指定列和列值,返回不含有指定列和列值的整行数据。
- 前缀过滤器按照具有特定前缀的行键来筛选数据。
- 列前缀过滤器按照列名的前缀(列簇名称)来筛选数据。
- 分页过滤器按照指定的页面行数来筛选数据,返回对应行数的结果集。

2. 过滤器的参数

过滤器的参数主要有比较运算符和比较器。HBase 提供了枚举类型的变量来表示抽象的比较运算符:LESS(<)、LESS_OR_EQUAL(<=)、EQUAL(=)、NOT_EQUAL(<>)、GREATER_OR_EQUAL(>=)、GREATER(>)、NO_OP(排除所有)。

8.6.2 比较器

比较器主要用于处理具体的比较逻辑，如字节级、字符串级的比较等。

1. 比较器种类

HBase 过滤器的常用比较器有如下几种。

（1）BinaryComparator 用于按字节索引顺序比较指定的字节数组。

（2）BinaryPrefixComparator 用于按字节索引顺序比较指定的字节数组，只比较指定的字节数组的长度。

（3）RegexStringComparator 用于使用正则表达式比较指定的字节数组，仅支持 EQUAL 和 NOT_EQUAL 两个比较运算符。

（4）SubstringComparator 用于判断指定的子串是否出现在指定的字节数组中，仅支持 EQUAL 和 NOT_EQUAL 两个比较运算符。

2. 比较器的语法

（1）比较器的一般语法如下。

```
ComparatorType:ComparatorValue
```

各种比较器的 ComparatorType：BinaryComparator 是 binary；BinaryPrefixComparator 是 binaryprefix；RegexStringComparator 是 regexstring；SubStringComparator 是 substring。ComparatorValue 可以是任意值。

（2）比较器示例。

binary:ab 匹配字典顺序大于 ab 的所有数据。

binaryprefix:ab 匹配前 2 个字符在字典上等于 ab 的所有数据。

regexstring:a*n 匹配以 a 开头且以 n 结尾的所有数据。

substring:ab12 匹配以字符串 ab12 开头的所有数据。

8.6.3 编程实例

1. 比较过滤器和比较器

【例 8-5】ComparedFilterTest.java

行过滤器的编程实例。

```
1    //比较过滤器
2    public class ComparedFilterTest {
3        //行过滤器
4        @Test
5        public void rowFilterTest() throws Exception {
6            Configuration conf = HBaseConfiguration.create();
7            conf.set("fs.defaultFS", "hdfs://192.168.142.131:8020");
8            conf.set("hbase.zookeeper.quorum", "qf01,qf02,qf03");
9            Connection conn = ConnectionFactory.createConnection(conf);
10           //通过表名称获取表的实例
11           Table table = conn.getTable(TableName.valueOf("t1"));
```

```
12          //创建扫描器
13          Scan scan = new Scan();
14          /*
15          //使用BinaryComparator过滤出Rowkey中比"row1"字节索引顺序靠后的行
16          RowFilter filter = new RowFilter(CompareFilter.CompareOp.GREATER, new BinaryComparator(Bytes.toBytes("row1")));
17          //使用SubstringComparator,过滤出Rowkey中所有含有"row"字符串的行
18          RowFilter filter = new RowFilter(CompareFilter.CompareOp.EQUAL, new SubstringComparator("row"));
19          */
20          /**
21           * RegexStringComparator解析
22           * .表示匹配任意单个字符
23           * '*'表示匹配前一个表达式一次或多次
24           * '.*'表示匹配任意字符或表达式
25           */
26          //使用RegexStringComparator过滤出Rowkey中所有含有".*w2"字符串的行
27          RowFilter filter = new RowFilter(CompareFilter.CompareOp.LESS_OR_EQUAL, new RegexStringComparator(".*w2"));
28          scan.setFilter(filter);
29          //通过表的扫描器得到结果集
30          ResultScanner rs = table.getScanner(scan);
31          //使用迭代器打印出数据
32          Iterator<Result> it = rs.iterator();
33          //调用HBaseTest类的printData()方法
34          HBaseTest.printData(it);
35          table.close();
36      }
```

运行结果如下。

```
row1,f1:id,1
row1,f2:age,22
row1,f2:id,2
row1,f2:name,jack
row2,f1:city,Beijing
```

列簇过滤器的编程实例。

```
1   //列簇过滤器
2   @Test
3   public void familyFilterTest() throws Exception {
4       Configuration conf = HBaseConfiguration.create();
5       conf.set("fs.defaultFS", "hdfs://192.168.142.131:8020");
6       conf.set("hbase.zookeeper.quorum", "qf01,qf02,qf03");
7       Connection conn = ConnectionFactory.createConnection(conf);
8       Table table = conn.getTable(TableName.valueOf("t1"));
9       Scan scan = new Scan();
10      //过滤出列簇中字节索引顺序大于"f1"的数据所在的行
11      FamilyFilter filter = new FamilyFilter(CompareFilter.CompareOp.GREATER, new BinaryComparator(Bytes.toBytes("f1")));
12      scan.setFilter(filter);
13      ResultScanner rs = table.getScanner(scan);
```

```
14          Iterator<Result> it = rs.iterator();
15          HBaseTest.printData(it);
16          table.close();
17      }
```

运行结果如下。

```
row1,f2:age,22
row1,f2:id,2
row1,f2:name,jack
```

列过滤器的编程实例。

```
1   //列过滤器
2       @Test
3       public void qualifierFilterTest() throws Exception {
4           Configuration conf = HBaseConfiguration.create();
5           conf.set("fs.defaultFS", "hdfs://192.168.142.131:8020");
6           conf.set("hbase.zookeeper.quorum", "qf01,qf02,qf03");
7           Connection conn = ConnectionFactory.createConnection(conf);
8           TableName tn = TableName.valueOf("t1");
9           HTableDescriptor htd = new HTableDescriptor(tn);
10          //添加列簇到HTableDescriptor
11          htd.addFamily(new HColumnDescriptor("f3"));
12          //向指定的Rowkey和列簇中添加列数据
13          Put data = new Put(Bytes.toBytes("row2"));
14          data.addColumn(Bytes.toBytes("f1"), Bytes.toBytes("name"), Bytes.toBytes("tom"));
15          Put data2 = new Put(Bytes.toBytes("row3"));
16          data2.addColumn(Bytes.toBytes("f1"), Bytes.toBytes("id"), Bytes.toBytes("3"));
17          data2.addColumn(Bytes.toBytes("f1"), Bytes.toBytes("name"), Bytes.toBytes("sophie"));
18          //添加数据到表中
19          Table table = conn.getTable(tn);
20          table.put(data);
21          table.put(data2);
22          //设定扫描的Rowkey范围(按字节索引排序)为: row1至row3(不包含row3)
23          Scan scan = new Scan(Bytes.toBytes("row1"), Bytes.toBytes("row3"));
24          //过滤出含有name列的行
25          QualifierFilter filter = new QualifierFilter(CompareFilter.CompareOp.EQUAL, new BinaryComparator(Bytes.toBytes("name")));
26          scan.setFilter(filter);
27          ResultScanner rs = table.getScanner(scan);
28          Iterator<Result> it = rs.iterator();
29          HBaseTest.printData(it);
30          table.close();
31      }
```

运行结果如下。

```
row1,f2:name,jack
row2,f1:name,tom
```

值过滤器的编程实例。

```
1    //值过滤器
2    @Test
3    public void valueFilterTest() throws Exception {
4        Configuration conf = HBaseConfiguration.create();
5        conf.set("fs.defaultFS", "hdfs://192.168.142.131:8020");
6        conf.set("hbase.zookeeper.quorum", "qf01,qf02,qf03");
7        Connection conn = ConnectionFactory.createConnection(conf);
8        Table table = conn.getTable(TableName.valueOf("t1"));
9        Scan scan = new Scan();
10       //使用RegexStringComparator过滤出列值为以2结尾的行
11       ValueFilter filter = new ValueFilter(CompareFilter.CompareOp.EQUAL, new RegexStringComparator(".*2"));
12       scan.setFilter(filter);
13       ResultScanner rs = table.getScanner(scan);
14       Iterator<Result> it = rs.iterator();
15       HBaseTest.printData(it);
16       table.close();
17   }
18   }
```

运行结果如下。

```
row1,f2:age,22
row1,f2:id,2
```

分别运行各个方法，即可验证运行结果。

2. 专用过滤器

【例8-6】DedicatedFilterTest.java

```
1    //专用过滤器之单列值过滤器
2    public class DedicatedFilterTest {
3        @Test
4        public void scvFilterTest() throws Exception {
5            Configuration conf = HBaseConfiguration.create();
6            conf.set("fs.defaultFS", "hdfs://192.168.142.131:8020");
7            conf.set("hbase.zookeeper.quorum", "qf01,qf02,qf03");
8            Connection conn = ConnectionFactory.createConnection(conf);
9            Table table = conn.getTable(TableName.valueOf("t1"));
10           Scan scan = new Scan();
11           //过滤出指定列和列值的行
12           SingleColumnValueFilter filter = new SingleColumnValueFilter(
13               Bytes.toBytes("f2"), Bytes.toBytes("name"),
14               CompareFilter.CompareOp.EQUAL,
15               new SubstringComparator("jack"));
16           //如果setFilterIfMissing()不设置为true,则那些不包含指定列和列值的行也会返回
17           filter.setFilterIfMissing(true);
18           scan.setFilter(filter);
19           ResultScanner rs = table.getScanner(scan);
20           Iterator<Result> it = rs.iterator();
21           HBaseTest.printData(it);
22           table.close();
23       }
24   }
```

运行结果如下。

```
row1,f1:id,1
row1,f2:age,22
row1,f2:id,2
row1,f2:name,jack
```

3. 过滤器组合使用

过滤器组合使用是指使用多个过滤器筛选返回到客户端的结果，需要使用 FilterList 类。

【例 8-7】 UnionFilterTest.java

```
1    //组合过滤器
2    public class UnionFilterTest {
3        @Test
4        public void testCombineFilter() throws Exception {
5            Configuration conf = HBaseConfiguration.create();
6            conf.set("fs.defaultFS", "hdfs://192.168.142.131:8020");
7            conf.set("hbase.zookeeper.quorum", "qf01,qf02,qf03");
8            Connection conn = ConnectionFactory.createConnection(conf);
9            Table table = conn.getTable(TableName.valueOf("t1"));
10           Scan scan = new Scan();
11           //使用列簇过滤器、BinaryComparator 过滤出含有 f1 列簇的行
12           FamilyFilter filter2 = new FamilyFilter(CompareFilter.CompareOp.EQUAL, new BinaryComparator(Bytes.toBytes("f1")));
13           //使用行过滤器、RegexStringComparator 过滤出 2 结尾的行
14           RowFilter filter1 = new RowFilter(CompareFilter.CompareOp.EQUAL, new RegexStringComparator(".*2"));
15           //使用单列值过滤器、BinaryComparator 过滤出指定列和列值的行
16           SingleColumnValueFilter filter3 = new SingleColumnValueFilter(Bytes.toBytes("f1"), Bytes.toBytes("age"),
17                   CompareFilter.CompareOp.GREATER, new BinaryComparator(Bytes.toBytes("23")));
18           filter3.setFilterIfMissing(true);
19           List<Filter> list = new ArrayList<Filter>();
20           list.add(filter1);
21           list.add(filter2);
22           list.add(filter3);
23           //至少有一个过滤器满足条件
24           FilterList fl = new FilterList(FilterList.Operator.MUST_PASS_ONE, list);
25           scan.setFilter(fl);
26           ResultScanner rs = table.getScanner(scan);
27           Iterator<Result> it = rs.iterator();
28           HBaseOperate.printData(it);
29           table.close();
30       }
31   }
```

运行结果如下。

```
row1,f1:id,1
row1,f2:age,22
row1,f2:id,2
row1,f2:name,jack
row2,f1:city,Beijing
```

```
row2,f1:name,tom
row3,f1:id,3
row3,f1:name,sophie
```

8.7　HBase 与 Hive 结合

8.7.1　HBase 与 Hive 结合的原因

HBase 和 Hive 都是 Hadoop 大数据生态圈中重要的组件，HBase 主要解决实时数据查询问题，Hive 主要解决批量数据离线处理问题。在使用 HBase 时，会遇到基于 HBase 表的复杂操作，进行 MapReduce 编程难度较大，其中一部分操作可以使用 Hive 提供的操作 HBase 表的接口来完成，为开发人员带来了一定的便利。需要注意的是，Hive 底层仍然调用的是 MapReduce，因此在性能上提升较少，开发人员需要根据实际情况酌情使用。

8.7.2　Hive 关联 HBase

1. 创建 Hive 表时关联新的 HBase 表

（1）在虚拟机 qf01 上启动 ZooKeeper 集群、Hadoop 集群、HBase、Hiveserver2 服务。

```
[root@qf01 ~]# xzk.sh start
[root@qf01 ~]# start-dfs.sh
[root@qf01 ~]# start-yarn.sh
[root@qf01 ~]# start-hbase.sh
[root@qf01 ~]# hiveserver2
```

（2）在虚拟机 qf01 上，重新打开一个终端，使用 Beeline 客户端连接 Hiveserver2 服务。

```
[root@qf01 ~]# beeline -u jdbc:hive2://localhost:10000 -n root
...
0: jdbc:hive2://localhost:10000>
```

（3）创建 Hive 表，同时建立 Hive 表与 HBase 表的映射关系。

```
jdbc:hive2://> create table hive_hbase(id string, name string, age string)
stored by 'org.apache.hadoop.hive.hbase.HBaseStorageHandler'
with serdeproperties("hbase.columns.mapping" = ":key,f1:name,f1:age")
tblproperties("hbase.table.name" = "hbase_hivetbl");
```

提示：

① stored by 用于指定存储处理器。hive_hbase 表使用的是 HBase 存储处理器。

② serdeproperties 用于指定 Hive 表和 HBase 表的映射关系。hive_hbase 表中的 id 对应 HBase 表中的行键，表示形式是:key；name 对应 f1:name；age 对应列 f1:age。其中 f1 是列簇。

③ tblproperties 用于指定 HBase 表的属性。hive_hbase 表中声明了 HBase 表的表名为 hbase_hivetbl。hbase.table.name 属性为可选项，如果未声明，默认与 hive_hbase 表同名。

（4）验证 Hive 的 hive_hbase 表是否创建成功。通过查看 hive_hbase 表的表结构来进行验证。

```
jdbc:hive2://> desc hive_hbase;
+-----------+------------+----------+--+
| col_name  | data_type  | comment  |
```

```
+-----------+-----------+----------+--+
| id        | string    |          |
| name      | string    |          |
| age       | string    |          |
+-----------+-----------+----------+--+
3 rows selected (0.231 seconds)
```

返回以上信息表明 hive_hbase 表创建成功。

（5）验证 HBase 的 hbase_hivetbl 表是否创建成功。

① 在虚拟机 qf01 上，重新打开一个终端，启动 HBase Shell 命令行。

```
[root@qf01 ~]# hbase shell
```

② 通过查看 hbase_hivetbl 表的表描述来进行验证。

```
hbase(main):003:0> desc 'hbase_hivetbl'
Table hbase_hivetbl is ENABLED
hbase_hivetbl
COLUMN FAMILIES DESCRIPTION
{NAME => 'f1', BLOOMFILTER => 'ROW', VERSIONS => '1', IN_MEMORY => 'false', KEEP_DELETED_CELLS => 'FALSE', D
ATA_BLOCK_ENCODING => 'NONE', TTL => 'FOREVER', COMPRESSION => 'NONE', MIN_VERSIONS => '0', BLOCKCACHE => 't
rue', BLOCKSIZE => '65536', REPLICATION_SCOPE => '0'}
1 row(s) in 0.3890 seconds
```

返回以上信息表明 hbase_hivetbl 表创建成功。至此，创建 Hive 表时关联新的 HBase 表完成。

2. 创建 Hive 表时关联已存在的 HBase 表

（1）创建 Hive 表 hive_hbase2，同时关联已存在的 HBase 表 ns1:t1。

```
jdbc:hive2://> create external table hive_hbase2(rowkey string, age string)
stored by 'org.apache.hadoop.hive.hbase.HBaseStorageHandler'
with serdeproperties("hbase.columns.mapping" = ":key, cf1:age")
tblproperties("hbase.table.name" = "ns1:t1");
```

注意：创建 Hive 表关联已存在的 HBase 表时，需要创建 Hive 外部表。

（2）查看 hive_hbase2 表的数据。

```
jdbc:hive2://> select * from hive_hbase2;
+---------------------+------------------+--+
| hive_hbase2.rowkey  | hive_hbase2.age  |
+---------------------+------------------+--+
| row1                | 18               |
+---------------------+------------------+--+
1 row selected (2.783 seconds)
```

返回以上信息，表明创建 Hive 表 hive_hbase2 时关联了 HBase 表 ns1:t1。在建立关联之后，可以实现使用 Hive 操作 HBase 表中的数据。

8.8 HBase 性能优化

为了高效地使用 HBase，需要对 HBase 进行性能优化。HBase 性能优化的常用方法如下。

1. API 性能优化

用户通过客户端使用 API 读写数据时，可以采用以下方法对 HBase 进行性能优化。

（1）关闭自动刷新写入。使用 setAutoFlush(false) 方法，关闭 HBase 表的自动刷新写入功能。向 HBase 表中大量写入数据（进行 put 操作）时，如果关闭自动刷新写入功能，写入的数据会先存放到一个缓冲区，等缓冲区被填满后再传送到 HRegion 服务器。如果启动了自动刷新写入功能，每进行一次 put 操作都会将数据传送到 HRegion 服务器，会增加网络负载。

（2）设置扫描范围。在使用扫描器处理大量的数据时，可以设置扫描指定的列数据，避免扫描未使用的数据，减少内存开销。

（3）关闭 ResultScanner。通过扫描器获取数据后，关闭 ResultScanner，这样可以尽快释放对应的 HRegionServer 的资源。

（4）使用过滤器。使用过滤器过滤出需要的数据，尽量减少服务器通过网络返回到客户端的数据量。

（5）批量写数据。通过调用 HTable.put(Put) 方法只能将一个指定的行键记录写入 HBase，而通过调用 HTable.put(List<Put>) 方法可以将指定的行键列表批量写入多行记录，减少网络开销。

2. 优化配置

（1）增加处理数据的线程数。在 /mysoft/hbase/conf/hbase-site.xml 文件中设置 HRegionServer 处理 I/O 请求的线程数，即设置 hbase.regionserver.handler.count，默认值为 10。

```
<property>
    <name>hbase.regionserver.handler.count</name>
    <value>10</value>
</property>
```

该线程数通常的设置范围为 100~200，可以提高 HRegionServer 的性能。需要注意的是，当数据量很大时，如果该值设置过大，则 HBase 处理的数据会占用较多的内存，因此该值不是越大越好。

（2）增加堆内存。在 /mysoft/hbase/conf/hbase-env.sh 文件中修改堆内存的大小，可以根据实际情况增加堆内存。

```
export HBASE_HEAPSIZE=1G            //默认值为 1GB
```

（3）调整 HRegion 的大小。在 /mysoft/hbase/conf/hbase-site.xml 文件中修改 HRegion 的大小。

```
<property>
    <name>hbase.hregion.max.filesize</name>
    <value>256MB</value>
</property>
```

通常，HBase 使用较小的 HRegion，可以使 HBase 集群更加平稳地运行；使用较大的 HRegion，能够减少 HBase 集群的 HRegion 数量。HBase 中 HRegion 的默认大小是 256MB，用户可以配置 1GB 以上的 HRegion。

（4）调整堆中块缓存大小。在 /mysoft/hbase/conf/hbase-site.xml 文件中修改块缓存大小。

```
<property>
    <name>perf.hfile.block.cache.size</name>
    <value>0.2</value>
</property>
```

该参数的默认值是 0.2。适当增大堆中块缓存可以提高 HBase 读取大量数据时的效率。

（5）调整 MemStore 的大小。在 /mysoft/hbase/conf/hbase-site.xml 文件中修改 MemStore 的大小。设置最大 MemStore，默认为堆内存的 40%（0.4）。

```xml
<property>
    <name>hbase.regionserver.global.memstore.size</name>
    <value>0.4</value>
</property>
```

设置最小 MemStore，默认为最大 MemStore 的 95%。

```xml
<property>
    <name>hbase.regionserver.global.memstore.size.lower.limit</name>
    <value>0.38</value>
</property>
```

8.9 本章小结

本章重点讲解了 HBase 架构、安装、HBase Shell 操作、HBase 编程、HBase 过滤器和比较器等 HBase 的核心内容，简单介绍了 HBase 与 Hive 两者结合的使用方法。通过对 HBase Shell 操作的实际演练，应深刻体会 HBase 的基本作用；通过学习 HBase 编程、HBase 优化，应深入理解 HBase 处理数据的高效性。

8.10 习题

1. 填空题

（1）HBase 利用 Hadoop 的_____作为其文件存储系统。
（2）HBase 的文件存储格式主要有两种：_____和_____。
（3）启动 HBase Shell 命令行的命令是_____。
（4）HBase 中行键保存为_____。
（5）HBase 主要涉及 4 个模块：_____、_____、_____和_____。

2. 选择题

（1）(   ) 不是 HBase 的特点。
    A. 大　　　　B. 稀疏　　　　C. 面向列　　　　D. 面向行
（2）列簇是 (   ) 的集合。
    A. 列　　　　B. Qualifier　　　C. Rowkey　　　　D. 值
（3）(   ) 是 HBase 存储的核心。
    A. Storage　　B. Store　　　　C. StoreFile　　　D. MemStore
（4）(   ) 主要存放用户建表时未指定命名空间的表。
    A. table　　　B. default　　　C. namespace　　　D. hbase
（5）以下属于比较过滤器的有 (   )。
    A. 行过滤器　　B. 列簇过滤器　　C. 列过滤器　　　D. 值过滤器

3. 思考题

（1）Hbase 的数据模型主要有哪些？
（2）简述 HBase 与 Hive 的区别。

第9章 Flume

本章学习目标
- 掌握 Flume 架构及其原理
- 熟悉 Flume 的安装和使用
- 掌握 Source、Sink、Channel 的使用方法
- 掌握拦截器的用法

要想实现对海量数据进行分析处理，首先需要将各种应用程序产生的海量数据高效地收集汇总，并传输到指定的数据存储区。Flume 作为高效的分布式数据采集工具应运而生。Flume 是一个基于流数据的简单而灵活的架构，用户通过给 Flume 添加各种新的功能来满足个性化的需求。

9.1 认识 Flume

9.1.1 Flume 简介

Flume 最初是 Cloudera 公司推出的日志采集系统，于 2009 年被捐赠给了 Apache 软件基金会，成为 Hadoop 相关组件之一。近几年随着 Flume 的不断完善、升级版本的推出，以及 Flume 内部各种组件的增加，用户在开发过程中使用 Flume 的便利性得到了很大的改善。

Flume 是一种可配置、高可用的数据采集工具，主要用于采集来自各种流媒体的数据（Web 服务器的日志数据等）并传输到集中式数据存储区域。Flume 支持在日志系统中定制各种数据发送方，用于收集数据；并且可以对数据进行简单处理，将其写到可定制的各种数据接受方（如文本、HDFS、HBase 等）。

Flume 有两个系列：Flume OG 和 Flume NG。Flume OG 是 Flume 0.9.x 系列，Flume NG 是 Flume 1.x 系列。目前使用 Flume NG 的企业较多，因此本书主要讲解 Flume NG。

9.1.2 Flume 的特点

Flume 的特点主要体现在以下几个方面。

（1）具有复杂的流动性。Flume 允许用户构建多跳流，允许使用扇入流和扇出流、上下文路由和故障跳转的备份路由（故障转移）。

① 多跳流。Flume 中可以有多个代理（Agent）。事件（Event）需要通过多个代理才能到达最终目的地，这样的数据流被称为多跳流。Flume 的数据流由事件贯穿始终。

② 扇出流（一对多形式）。从一个源（Source）到多个通道（Channel）的数据流被称为扇出流。

③ 扇入流（多对一形式）。从多个源到一个通道的数据流被称为扇入流。

（2）具有可靠性。Flume 的源和接收器（Sink）分别封装在事务中，可以确保事件集在数据流中从一个点到另一个点进行可靠的传递。

（3）具有可恢复性。事件存储在通道中，当 Flume 出现故障时，通道负责恢复数据。

9.2 Flume 基本组件

Flume 基本组件主要包括 Event 和 Agent。

9.2.1 Event

Event 是 Flume 中具有有效负载的字节数据流和可选的字符串属性集，是 Flume 传送数据的基本单位。Event 由 Header 和 Body 组成。Header 是一个 Map<String，String>，存储字符串属性集；Body 是一个字节数组，存储字节数据。

9.2.2 Agent

Agent 是一个虚拟机进程，负责将外部来源产生的消息转发到外部目的地。Agent 由 Source、Channel 和 Sink 构成。

1. Source

Source 从外部来源读入 Event，并写入 Channel。每个 Source 可以发送 Event 到多个 Channel 中。Source 的常见类型如表 9.1 所示。

表 9.1　　　　　　　　　　　　　　　　Source 的常见类型

类型	简介
Netcat Source	监控某个端口，读取流经端口的每一个文本行数据
Exec Source	Source 启动的时候会运行一个设置好的 Linux 命令，该命令不断地往标准输出（stdout）中输出数据，这些数据会被打包成 Event 进行处理
Spooling Directory Source	监听指定目录，当该目录有新文件出现时，把文件的数据打包成 Event 进行处理
Syslog Source	读取 Syslog 数据，产生 Event，支持 UDP 和 TCP 两种协议
Stress Source	用户可以配置要发送的事件总数以及要传递的最大事件数，多用于负载测试
HTTP Source	基于 HTTP POST 或 GET 方式的数据源，支持 JSON、BLOB 表示形式
Avro Source	支持 Avro RPC 协议，提供一个 Avro 的接口，往设置的地址和端口发送 Avro 消息，Source 就能接收到，例如，Log4j Appender 通过 Avro Source 将消息发送到 Agent
Taildir Source	监听实时追加内容的文件
Thrift Source	支持 Thrift 协议，提供一个 Thrift 接口，类似 Avro
JMS Source	从 Java 消息服务读取数据

2. Sink

Sink 从 Channel 中读取 Event 后写入目的地。每个 Sink 只能从一个 Channel 中获取数据。Sink 的常见类型如表 9.2 所示。

表 9.2　　　　　　　　　　　　　　　Sink 的常见类型

类型	简介
HDFS Sink	将数据写入 HDFS，默认写入格式为 SequenceFile
Logger Sink	将数据写入日志文件
Hive Sink	将数据写入 Hive
File Roll Sink	将数据存储到本地文件系统，多用作数据收集
HBase Sink	将数据写入 HBase
Thrift Sink	将数据转换成 Thrift Event 后，发送到配置的 RPC 端口上
Avro Sink	将数据转换成 Avro Event 后，发送到配置的 RPC 端口上
Null Sink	丢弃所有数据
ElasticSearch Sink	将数据发送到 ElasticSearch 集群
Kite Dataset Sink	写数据到 Kite Dataset，试验性质

3. Channel

Channel 相当于数据缓冲区，接收 Source 传入的事件后发送给 Sink。Channel 的常见类型如表 9.3 所示。

表 9.3　　　　　　　　　　　　　　Channel 的常见类型

类型	简介
Memory Channel	数据存储在内存中，可以实现高速的数据吞吐，Flume 出现故障时，数据会丢失
File Channel	数据存储在磁盘文件中，可以持久化所有的 Event，Flume 出现故障时，数据不会丢失

9.3　Flume 安装

本书使用的是 1.8.0 版本的 Flume，安装包见附录。Flume 的安装步骤如下。

（1）将 Flume 安装包 apache-flume-1.8.0-bin.tar.gz 放到虚拟机 qf01 的/root/Downloads/目录下，切换到 root 用户，解压 Flume 安装包到/mysoft 目录下。

```
[root@qf01 ~]# tar -zxvf /root/Downloads/apache-flume-1.8.0-bin.tar.gz -C /mysoft/
```

（2）切换到/mysoft 目录下，将 apache-flume-1.8.0-bin 重命名为 flume。

```
[root@qf01 ~]# cd /mysoft/
[root@qf01 mysoft]# mv apache-flume-1.8.0-bin flume
```

（3）打开/etc/profile 文件，配置 Flume 环境变量。

```
[root@qf01 mysoft]# vi /etc/profile
```

在文件末尾添加如下三行内容。

```
# Flume environment variables
export FLUME_HOME=/mysoft/flume
export PATH=$PATH:$FLUME_HOME/bin
```

（4）使环境变量生效。

```
[root@qf01 mysoft]# source /etc/profile
```

（5）修改 Flume 的配置文件。

① 切换到/mysoft/flume/conf 目录下。

```
[root@qf01 mysoft]# cd /mysoft/flume/conf
```

② 将 flume-env.ps1.template 重命名为 flume-env.ps1。

```
[root@qf01 conf]# mv flume-env.ps1.template flume-env.ps1
```

③ 将 flume-env.sh.template 重命名为 flume-env.sh。

```
[root@qf01 conf]# mv flume-env.sh.template flume-env.sh
```

④ 修改 flume-env.sh，将# export JAVA_HOME=/usr/lib/jvm/java-8-oracle 一行替换为如下内容。

```
export JAVA_HOME=/usr/java/jdk1.8.0_121
```

（6）查看 Flume 的版本信息。

```
[root@qf01 conf]# flume-ng version
Flume 1.8.0
```

至此，Flume 安装完成。

9.4 Flume 数据流模型

Flume 常见的数据流模型有单 Agent 数据流模型、多 Agent 串联数据流模型、多 Agent 汇集数据流模型和单 Agent 多路数据流模型。

1. 单 Agent 数据流模型

单 Agent 数据流模型，如图 9.1 所示。

在图 9.1 中，一个 Agent 由一个 Source、一个 Channel、一个 Sink 构成。

图 9.1 单 Agent 数据流模型

2. 多 Agent 串联数据流模型

多 Agent 串联数据流模型，如图 9.2 所示。

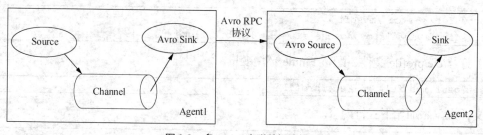

图 9.2 多 Agent 串联数据流模型

在图 9.2 中，为了使数据在多个 Agent 中流通，Agent1 中的 Sink 和 Agent2 中的 Source 需要是 Avro 类型，Agent1 中的 Sink 指向 Agent2 中 Source 的主机名（或 IP 地址）和端口。

3. 多 Agent 汇集数据流模型

多 Agent 汇集数据流模型是采集大量日志数据时常用的数据流模型，如图 9.3 所示。例如，将从数百个 Web 服务器采集的日志数据发送给写入 HDFS 集群的十几个 Agent，就采用该模型。

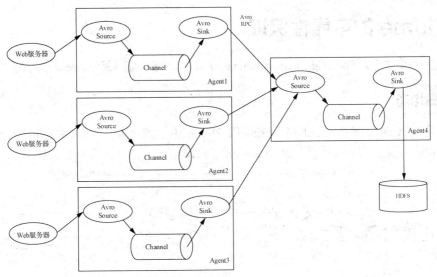

图 9.3　多 Agent 汇集数据流模型

4. 单 Agent 多路数据流模型

单 Agent 多路数据流模型，如图 9.4 所示。

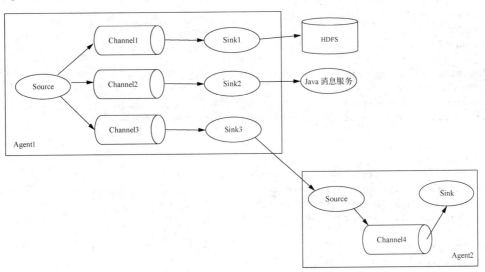

图 9.4　单 Agent 多路数据流模型

在图 9.4 中，Agent1 由一个 Source、多个 Channel、多个 Sink 构成。一个 Source 接收 Event，并将 Event 发送到三个 Channel 中，Channel 对应的 Sink 处理各 Channel 内的 Event，然后将数据分别存储到指定的位置。

Source 将 Event 发送到 Channel 中可以采用两种不同的策略：Replicating（复制通道选择器）和 Multiplexing（多路复用通道选择器）。

（1）Replicating 是 Source 将每个 Event 发送到每个与它连接的 Channel 中，也就是将 Event 复制多份发到不同的 Channel。

（2）Multiplexing 是 Source 根据 Header 中的某一个键的值决定将 Event 发送到哪个 Channel 中。

9.5 Flume 的可靠性保证

Flume 的一些组件（如 Spooling Directory Source、File Channel）能够保证 Agent 出问题后不丢失数据。

9.5.1 负载均衡

Source 里的 Event 流经 Channel，进入 Sink 组，在 Sink 组内部根据负载算法（round_robin、random）选择 Sink，后续可以选择不同机器上的 Agent 实现负载均衡，如图 9.5 所示。

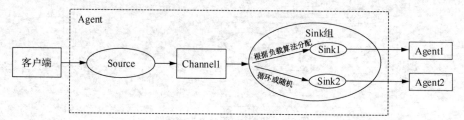

图 9.5 负载均衡

具体配置如下。

```
# Name the components on this agent
a1.sources = r1
a1.sinks = k1 k2
a1.channels = c1

# Describe/configure the source
a1.sources.r1.type = exec
a1.sources.r1.channels=c1
a1.sources.r1.command=tail -F /home/flume/xx.log

#define sinkgroups
a1.sinkgroups=g1
a1.sinkgroups.g1.sinks=k1 k2
a1.sinkgroups.g1.processor.type=load_balance
a1.sinkgroups.g1.processor.backoff=true
a1.sinkgroups.g1.processor.selector=round_robin

#define the sink 1
a1.sinks.k1.type=avro
a1.sinks.k1.hostname=192.168.1.112
a1.sinks.k1.port=9876

#define the sink 2
a1.sinks.k2.type=avro
a1.sinks.k2.hostname=192.168.1.113
a1.sinks.k2.port=9876

# Use a channel which buffers events in memory
a1.channels.c1.type = memory
a1.channels.c1.capacity = 1000
a1.channels.c1.transactionCapacity = 100

# Bind the source and sink to the channel
a1.sources.r1.channels = c1
a1.sinks.k1.channel = c1
```

```
a1.sinks.k2.channel=c1
```

不同的 Agent 处理同一个客户端产生的数据，实现负载均衡。

```
log4j.rootLogger=INFO,flume
log4j.appender.flume = org.apache.flume.clients.log4jappender.LoadBalancingLog4jAppender
log4j.appender.flume.Hosts = 192.168.1.111:41414 192.168.1.111:41414
```

9.5.2 故障转移

配置一组 Sink，这组 Sink 组成一个 Sink 故障转移处理器，当有一个 Sink 处理失败，Flume 将这个 Sink 放到一个地方，设定冷却时间，待其可以正常处理 Event 时再取回。

Event 通过一个 Channel 流向一个 Sink 组，在 Sink 组内部根据优先级选择具体的 Sink，失败后再转向另一个 Sink，流程如图 9.6 所示。

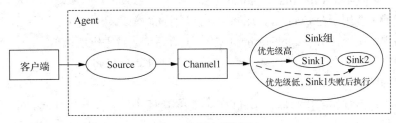

图 9.6 故障转移流程图

具体配置如下。

```
# Name the components on this agent
a1.sources = r1
a1.sinks = k1 k2
a1.channels = c1

# Describe/configure the source
a1.sources.r1.type = exec
a1.sources.r1.channels=c1
a1.sources.r1.command=tail -F /home/flume/xx.log

#define sinkgroups
a1.sinkgroups=g1
a1.sinkgroups.g1.sinks=k1 k2
a1.sinkgroups.g1.processor.type=failover
a1.sinkgroups.g1.processor.priority.k1=10
a1.sinkgroups.g1.processor.priority.k2=5
a1.sinkgroups.g1.processor.maxpenalty=10000

#define the sink 1
a1.sinks.k1.type=avro
a1.sinks.k1.hostname=192.168.1.112
a1.sinks.k1.port=9876

#define the sink 2
a1.sinks.k2.type=avro
a1.sinks.k2.hostname=192.168.1.113
a1.sinks.k2.port=9876

# Use a channel which buffers events in memory
```

```
a1.channels.c1.type = memory
a1.channels.c1.capacity = 1000
a1.channels.c1.transactionCapacity = 100

# Bind the source and sink to the channel
a1.sources.r1.channels = c1
a1.sinks.k1.channel = c1
a1.sinks.k2.channel=c1
```

9.6 Flume 拦截器

拦截器负责修改和删除 Event，每个 Source 可以配置多个拦截器，形成拦截器链（Interceptor Chain）。Flume 自带的拦截器主要有时间戳拦截器（Timestamp Interceptor）、主机拦截器（Host Interceptor）、静态拦截器（Static Interceptor）。

时间戳拦截器主要用于在 Event 的 Header 中加入时间戳。

主机拦截器主要用于在 Event 的 Header 中加入主机名（或者主机地址）。

静态拦截器主要用于在 Event 的 Header 中加入固定的 Key 和 Value。

1. 时间戳拦截器

时间戳拦截器是最常用的拦截器，下面以时间戳拦截器为例，演示拦截器的使用方法。

（1）在虚拟机 qf01 上，切换到/mysoft/flume/conf 目录下，新建 Flume 配置文件 interceptor_time.conf。

```
[root@qf01 conf]# vi interceptor_time.conf
```

添加以下内容。

```
# 命名Agent
a1.sources = r1
a1.sinks = k1
a1.channels = c1
# 配置Source
a1.sources.r1.type = netcat
a1.sources.r1.bind = localhost
a1.sources.r1.port = 8888
# 命名拦截器
a1.sources.r1.interceptors = i1
# 指定拦截器类型
a1.sources.r1.interceptors.i1.type = timestamp
# 配置Sink
a1.sinks.k1.type = logger
# 配置Channel
a1.channels.c1.type = memory
a1.channels.c1.capacity = 1000
a1.channels.c1.transactionCapacity = 100
# 配置Source 和Sink 使用的Channel
a1.sources.r1.channels = c1
a1.sinks.k1.channel = c1
```

（2）启动 Flume Agent a1。

```
[root@qf01 conf]# flume-ng agent -n a1 -f interceptor_time.conf
```

```
...
18/10/22 09:49:58 INFO source.NetcatSource: Created serverSocket:sun.nio.ch.Server
SocketChannelImpl[/127.0.0.1:8888]
```

（3）在虚拟机 qf01 上，重新打开一个终端，启动 Netcat 客户端，输入 hi xiaofeng 后，按回车键。

```
[root@qf01 ~]# nc localhost 8888
hi xiaofeng
OK
```

（4）查看启动 Flume Agent a1 的终端，出现以下内容。

```
18/10/22 09:51:40 INFO sink.LoggerSink: Event: { headers:{timestamp=1540173099879}
body: 68 69 20 78 69 61 6F 66 65 6E 67          hi xiaofeng }
```

由以上内容可知，Flume 通过时间戳拦截器在 Event 的 Header 中加入了时间戳 timestamp=1540173099879。

2. 拦截器链

（1）在虚拟机 qf01 上，切换到/mysoft/flume/conf 目录下，新建 Flume 配置文件 interceptor_chain.conf。

```
[root@qf01 conf]# vi interceptor_chain.conf
```

添加以下内容。

```
# 命名Agent
a1.sources = r1
a1.sinks = k1
a1.channels = c1
# 配置Source
a1.sources.r1.type = netcat
a1.sources.r1.bind = localhost
a1.sources.r1.port = 8888
# 命名拦截器链
a1.sources.r1.interceptors = i1 i2 i3
# 分别指定拦截器类型
a1.sources.r1.interceptors.i1.type = timestamp
a1.sources.r1.interceptors.i2.type = host
a1.sources.r1.interceptors.i3.type = static
a1.sources.r1.interceptors.i3.key = city
a1.sources.r1.interceptors.i3.value = Beijing
# 配置Sink
a1.sinks.k1.type = logger
# 配置Channel
a1.channels.c1.type = memory
a1.channels.c1.capacity = 1000
a1.channels.c1.transactionCapacity = 100
# 配置Source 和 Sink 使用的 Channel
a1.sources.r1.channels = c1
a1.sinks.k1.channel = c1
```

（2）启动 Flume Agent a1。

```
[root@qf01 conf]# flume-ng agent -n a1 -f interceptor_chain.conf
...
```

```
18/10/22 10:08:35 INFO source.NetcatSource: Created serverSocket:sun.nio.ch.ServerSocket
ChannelImpl[/127.0.0.1:8888]
```

(3)在虚拟机 qf01 上,重新打开一个终端,启动 Netcat 客户端,输入 hello 后,按回车键。

```
[root@qf01 ~]# nc localhost 8888
hello
OK
```

(4)查看启动 Flume Agent a1 的终端,出现以下内容。

```
18/10/22 10:09:05 INFO sink.LoggerSink: Event: { headers:{city=Beijing, host=192.168.
142.131, timestamp=1540174145729} body: 68 65 6C 6C 6F
```

由以上内容可知,Flume 通过 3 个拦截器在 Event 的 Header 中加入了指定的相关内容。

9.7 采集案例

9.7.1 采集目录到 HDFS

采集需求:某服务器的某特定目录下,会不断产生新的文件,每当有新文件出现,就需要把文件采集到 HDFS 中去。

根据需求,首先定义以下 3 大要素。

(1)Source:监控文件目录 spooldir。spooldir 有以下特性。

① 监视一个目录,只要目录中出现新文件,就采集文件中的内容。

② 采集完成的文件,会被 Agent 自动添加一个后缀:COMPLETED。

③ 所监视的目录中不允许出现文件名相同的文件。

(2)Sink:类型为 HDFS Sink。

(3)Channel:可以用 File Channel,也可以用 Memory Channel。

配置文件编写如下。

```
#定义三大组件的名称
agent1.sources = source1
agent1.sinks = sink1
agent1.channels = channel1
# 配置 Source 组件
agent1.sources.source1.type = spooldir
agent1.sources.source1.spoolDir = /home/hadoop/logs/
agent1.sources.source1.fileHeader = false
#配置拦截器
agent1.sources.source1.interceptors = i1
agent1.sources.source1.interceptors.i1.type = host
agent1.sources.source1.interceptors.i1.hostHeader = hostname
# 配置 Sink 组件
agent1.sinks.sink1.type = hdfs
agent1.sinks.sink1.hdfs.path =hdfs://hdp01:9000/weblog/flume-collection/%y-%m-%d/%H-%M
agent1.sinks.sink1.hdfs.filePrefix = access_log
agent1.sinks.sink1.hdfs.maxOpenFiles = 5000
agent1.sinks.sink1.hdfs.batchSize= 100
agent1.sinks.sink1.hdfs.fileType = DataStream
agent1.sinks.sink1.hdfs.writeFormat =Text
agent1.sinks.sink1.hdfs.rollSize = 102400
agent1.sinks.sink1.hdfs.rollCount = 1000000
```

```
agent1.sinks.sink1.hdfs.rollInterval = 60
#agent1.sinks.sink1.hdfs.round = true
#agent1.sinks.sink1.hdfs.roundValue = 10
#agent1.sinks.sink1.hdfs.roundUnit = minute
agent1.sinks.sink1.hdfs.useLocalTimeStamp = true
# Use a channel which buffers events in memory
agent1.channels.channel1.type = memory
agent1.channels.channel1.keep-alive = 120
agent1.channels.channel1.capacity = 500000
agent1.channels.channel1.transactionCapacity = 600
# Bind the source and sink to the channel
agent1.sources.source1.channels = channel1
agent1.sinks.sink1.channel = channel1
```

Channel 参数解释如下。

capacity：默认该通道中可以存储的 Event 的最大数量。

trasactionCapacity：每次可以从 Source 中拿到或者送到 Sink 中的 Event 的最大数量。

keep-alive：Event 添加到通道中或者移出的允许时间。

9.7.2 采集文件到 HDFS

采集需求：业务系统使用 Log4j 生成的日志，日志内容不断增加，需要把追加到日志文件中的数据实时采集到 HDFS。

根据需求，首先定义以下 3 大要素。

（1）Source：Exec Source 监控文件 tail -F file。

（2）Sink：类型为 HDFS Sink。

（3）Channel：可以用 File Channel，也可以用 Memory Channel。

配置文件编写如下。

```
agent1.sources = source1
agent1.sinks = sink1
agent1.channels = channel1

# Describe/configure tail -F source1
agent1.sources.source1.type = exec
agent1.sources.source1.command = tail -F /home/hadoop/logs/access_log
agent1.sources.source1.channels = channel1

#configure host for source
agent1.sources.source1.interceptors = i1
agent1.sources.source1.interceptors.i1.type = host
agent1.sources.source1.interceptors.i1.hostHeader = hostname

# Describe sink1
agent1.sinks.sink1.type = hdfs
#a1.sinks.k1.channel = c1
agent1.sinks.sink1.hdfs.path
=hdfs://hdp-node-01:9000/weblog/flume-collection/%y-%m-%d/%H-%M
agent1.sinks.sink1.hdfs.filePrefix = access_log
agent1.sinks.sink1.hdfs.maxOpenFiles = 5000
agent1.sinks.sink1.hdfs.batchSize= 100
agent1.sinks.sink1.hdfs.fileType = DataStream
agent1.sinks.sink1.hdfs.writeFormat =Text
agent1.sinks.sink1.hdfs.rollSize = 102400
agent1.sinks.sink1.hdfs.rollCount = 1000000
```

```
agent1.sinks.sink1.hdfs.rollInterval = 60
agent1.sinks.sink1.hdfs.round = true
agent1.sinks.sink1.hdfs.roundValue = 10
agent1.sinks.sink1.hdfs.roundUnit = minute
agent1.sinks.sink1.hdfs.useLocalTimeStamp = true

# Use a channel which buffers events in memory
agent1.channels.channel1.type = memory
agent1.channels.channel1.keep-alive = 120
agent1.channels.channel1.capacity = 500000
agent1.channels.channel1.transactionCapacity = 600

# Bind the source and sink to the channel
agent1.sources.source1.channels = channel1
agent1.sinks.sink1.channel = channel1
```

9.8 本章小结

本章对 Flume 的安装、数据流模型、Source、Sink、Channel、拦截器、通道选择器进行了讲解。通过 Source、Sink、Channel 的使用，读者可以深刻理解 Flume 的应用场景。通过编程，读者可以简单了解 Flume 的实际开发工作。

9.9 习题

1. 填空题

（1）_____是 Flume 传送数据的基本单位。
（2）Agent 由_____、_____和_____构成。
（3）Flume 自带的通道选择器主要有_____和_____。
（4）复制通道选择器主要用于从一个 Source 复制 Event 到多个_____中。
（5）Flume 是_____提供的一个高可用、高可靠、分布式的海量日志采集、聚合和传输系统。

2. 选择题

（1）Flume 是一种可配置、高可用的（   ）。
   A. 数据采集工具　B. 数据挖掘工具　C. 数据驱动工具　D. 数据可视化工具
（2）Event 由（   ）和（   ）组成。
   A. Header　　B. Body　　C. Leg　　D. Arm
（3）Flume 中常见的 Source 有（   ）。
   A. Netcat Source　　　　B. Exec Source
   C. Spooling Directory Source　　D. Syslog Source
（4）Flume 中常见的 Sink 有（   ）。
   A. HDFS Sink　B. File Roll Sink　C. Hive Sink　D. HBase Sink
（5）Flume 以（   ）为最小的独立运行单位。
   A. Stage　　B. Agent　　C. Task　　D. Job

3. 思考题

（1）Flume 数据流模型有哪些？
（2）Flume 自带的拦截器有哪些？

第10章 Sqoop

本章学习目标
- 了解 Sqoop 的原理及其安装
- 熟悉 Sqoop 的架构
- 掌握 Sqoop 的 import、export、job 命令的用法

Sqoop 通过 Hadoop 的 MapReduce 实现了数据在关系型数据库与 HDFS、Hive、HBase 等组件之间的传输。在大数据项目中，Sqoop 为大规模数据的处理与存储提供了重要支持。

10.1 认识 Sqoop

10.1.1 Sqoop 简介

Sqoop 是一种用于在 Hadoop 和结构化数据系统（如关系型数据库、大型机）之间高效传输数据的工具。Sqoop 项目开始于 2009 年，它的出现主要是为了满足以下两种需求。

（1）企业的业务数据大多存放在关系型数据库（如 MySQL、Oracle）中，数据量达到一定规模后，如果需要对其进行统计和分析，直接使用关系型数据库处理数据效率较低，这时可以通过 Sqoop 将数据从关系型数据库导入 Hadoop 的 HDFS（或 HBase、Hive）进行离线分析。

（2）用 Hadoop 处理后的数据，往往需要同步到关系型数据库中作为业务辅助数据，这时可以通过 Sqoop 将 Hadoop 中的数据导出到关系型数据库。

Sqoop 担负了将数据导入和导出 Hadoop 的任务。Sqoop 的核心设计思想是利用 MapReduce 提高数据传输速度。Sqoop 的导入和导出功能就是通过 MapReduce 作业来实现的。

目前 Sqoop 主要有两个系列：Sqoop 1 和 Sqoop 2。Sqoop 1 最新的稳定版本是 1.4.7，Sqoop 2 的最新版本是 1.99.7。1.99.7 版本功能不完整，并且与 1.4.7 版本不兼容，不适用于生产部署，目前大多数企业主要使用的还是 Sqoop 1，因此本书选用 1.4.7 版本进行讲解。

10.1.2 Sqoop 原理

Sqoop 的原理其实就是将导入导出命令转化为 MapReduce 程序来执行，Sqoop 在接收到命令后，都要生成 MapReduce 程序。

使用 Sqoop 的代码生成工具可以方便地查看到 Sqoop 生成的 Java 代码，并可在此基础之上进行深入定制开发。

10.1.3 Sqoop 架构

Sqoop 的常用架构如图 10.1 所示。

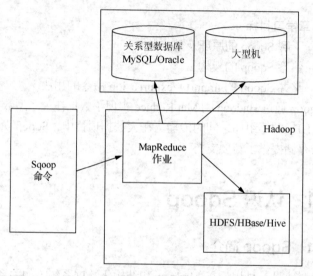

图 10.1　Sqoop 架构图

通过 Sqoop 架构图可以简单地了解 Sqoop 的运行流程。以从关系型数据库导入数据为例，Sqoop 大致的运行流程是：读取将要导入数据的表结构；生成运行类（默认是 QueryResult）；打成 jar 包；提交给 Hadoop；设置 MapReduce 作业的各种参数；由 Hadoop 来执行 MapReduce 作业。

10.2　Sqoop 安装

Sqoop 的安装包见附录。注意：Sqoop 的 1.4.7 版本底层适配的是 Hadoop 2.6.0 版本，本书使用的是 Hadoop 2.7.3 版本。虽然使用的 Hadoop 版本不同，但是 Hadoop 的 2.7.3 版本兼容 2.6.0 版本。Sqoop 的安装步骤如下。

（1）启动 ZooKeeper 集群、Hadoop 集群、MySQL。

```
[root@qf01 ~]# xzk.sh start
[root@qf01 ~]# start-dfs.sh
[root@qf01 ~]# start-yarn.sh
[root@qf01 ~]# mysql -uroot -p
Enter password:
...
mysql>
```

提示：当出现 Enter password 时输入 root，按回车键。

（2）将 Sqoop 安装包 sqoop-1.4.7.bin__hadoop-2.6.0.tar.gz 放到虚拟机 qf01 的/root/Downloads/目录下。在虚拟机 qf01 上，重新打开一个终端，解压 Sqoop 安装包到/mysoft 目录下。

```
[root@qf01 ~ ]# tar -zxvf /root/Downloads/sqoop-1.4.7.bin__hadoop-2.6.0.tar.gz -C /mysoft/
```

（3）切换到/mysoft 目录下，将 sqoop-1.4.7.bin__hadoop-2.6.0 重命名为 sqoop。

```
[root@qf01 ~]# cd /mysoft/
[root@qf01 mysoft]# mv sqoop-1.4.7.bin__hadoop-2.6.0 sqoop
```

（4）打开/etc/profile 文件，配置 Sqoop 环境变量。

```
[root@qf01 mysoft]# vi /etc/profile
```

在文件末尾添加如下三行内容。

```
# Sqoop environment variables
export SQOOP_HOME=/mysoft/sqoop
export PATH=$PATH:$SQOOP_HOME/bin
```

（5）使环境变量生效。

```
[root@qf01 mysoft]# source /etc/profile
```

（6）将/root/apps 目录下的 MySQL 的驱动文件 mysql-connector-java-5.1.41.jar（见附录）复制到虚拟机 qf01 的/mysoft/sqoop/lib 目录下。

```
[root@qf01 mysoft]# cp /root/Downloads/mysql-connector-java-5.1.41.jar /mysoft/sqoop/lib
```

（7）修改 Sqoop 的配置文件。

① 切换到/mysoft/sqoop/conf 目录下，将文件 sqoop-env-template.sh 重命名为 sqoop-env.sh。

```
[root@qf01 mysoft]# cd /mysoft/sqoop/conf
[root@qf01 conf]# mv sqoop-env-template.sh sqoop-env.sh
```

② 修改 sqoop-env.sh 文件。

```
[root@qf01 conf]# vi sqoop-env.sh
```

找到 sqoop-env.sh 文件的如下几行内容。

```
#export HADOOP_COMMON_HOME=
#export HADOOP_MAPRED_HOME=
#export HBASE_HOME=
#export HIVE_HOME=
#export ZOOCFGDIR=
```

将其分别替换为如下内容。

```
export HADOOP_COMMON_HOME=/usr/local/hadoop-2.7.3
export HADOOP_MAPRED_HOME=/usr/local/hadoop-2.7.3
export HBASE_HOME=/mysoft/hbase
export HIVE_HOME=/mysoft/hive
export ZOOCFGDIR=/mysoft/zookeeper/conf
```

（8）查看当前 Sqoop 的版本信息。

```
[root@qf01 conf]# sqoop version
Warning: /mysoft/sqoop/../hcatalog does not exist! HCatalog jobs will fail.
Please set $HCAT_HOME to the root of your HCatalog installation.
Warning: /mysoft/sqoop/../accumulo does not exist! Accumulo imports will fail.
Please set $ACCUMULO_HOME to the root of your Accumulo installation.
18/10/26 13:45:04 INFO sqoop.Sqoop: Running Sqoop version: 1.4.7
Sqoop 1.4.7
git commit id 2328971411f57f0cb683dfb79d19d4d19d185dd8
Compiled by maugli on Thu Dec 21 15:59:58 STD 2017
```

出现 Sqoop 1.4.7 一行，表明 Sqoop 安装成功。

（9）去除警告信息。

因为没有设置 HCAT_HOME 和 ACCUMULO_HOME，所以出现了相关的警告信息，可进行如下操作去除警告信息。（本书对 HCatalog 和 Accumulo 不做介绍，感兴趣的读者可以自行了解。）

① 修改/mysoft/sqoop/bin 目录下的 configure-sqoop 文件。

```
[root@qf01 conf]# cd /mysoft/sqoop/bin
[root@qf01 bin]# vi configure-sqoop
```

使用#注释掉 HCAT_HOME 和 ACCUMULO_HOME 的相关信息。

```
#if [ ! -d "${HCAT_HOME}" ]; then
#  echo "Warning: $HCAT_HOME does not exist! HCatalog jobs will fail."
#  echo 'Please set $HCAT_HOME to the root of your HCatalog installation.'
#fi
#if [ ! -d "${ACCUMULO_HOME}" ]; then
#  echo "Warning: $ACCUMULO_HOME does not exist! Accumulo imports will fail."
#  echo 'Please set $ACCUMULO_HOME to the root of your Accumulo installation.'
#fi
```

② 查看 Sqoop 的版本信息，警告信息已去除。

```
[root@qf01 bin]# sqoop version
18/10/26 13:50:20 INFO sqoop.Sqoop: Running Sqoop version: 1.4.7
Sqoop 1.4.7
git commit id 2328971411f57f0cb683dfb79d19d4d19d185dd8
Compiled by maugli on Thu Dec 21 15:59:58 STD 2017
```

至此，Sqoop 安装完成。

10.3 Sqoop 命令

Sqoop 主要通过命令（Command）来进行相关操作，Sqoop 命令是学习 Sqoop 的核心内容。可使用 help 命令查看 Sqoop 命令。Sqoop 常用命令如表 10.1 所示。

表 10.1　　　　　　　　　　　　　Sqoop 常用命令

序号	命令	对应的 Java 类	用途
1	import	ImportTool	将数据导入到集群
2	export	ExportTool	将集群数据导出
3	codegen	CodeGenTool	获取数据库中某张表的数据，生成 Java 类
4	create-hive-table	CreateHiveTableTool	创建 Hive 表
5	eval	EvalSqlTool	查看 SQL 语句的执行结果
6	import-all-tables	ImportAllTablesTool	将某个数据库下所有表导入 HDFS

续表

序号	命令	对应的 Java 类	用途
7	list-databases	ListDatabasesTool	列出 MySQL 的所有数据库名
8	list-tables	ListTablesTool	列出某个数据库下所有表
9	help	HelpTool	查看帮助信息
10	version	VersionTool	查看版本信息

10.3.1 Sqoop 数据库连接参数

Sqoop 数据库连接参数如表 10.2 所示。

表 10.2　　　　　　　　　　Sqoop 数据库连接参数

参数	说明
--connect	Java 数据库连接关系型数据库
--connection-manager	指定要使用的连接管理类
--driver	Java 数据库连接
--help	打印帮助信息
--password	连接数据库的密码
--username	连接数据库的用户名
--verbose	在控制台打印出详细信息

10.3.2 Sqoop export 参数

Sqoop export 命令的控制参数如表 10.3 所示。

表 10.3　　　　　　　　　Sqoop export 命令的控制参数

参数	说明
--input-enclosed-by char	给字段前后加上指定字符
--input-escaped-by char	对含有转义符的字段做转义处理
--input-fields-terminated-by char	字段之间的分隔符
--input-lines-terminated-by char	行之间的分隔符
--input-optionally-enclosed-by char	给带有双引号或单引号的字段前后加上指定字符

10.3.3 Sqoop import 参数

Sqoop import 命令的控制参数如表 10.4 所示。

表 10.4　　　　　　　　　Sqoop import 命令的控制参数

参数	说明
--enclosed-by char	给字段前后加上指定字符
--escaped-by char	对字段中的双引号加转义符
--fields-terminated-by char	设定每个字段以什么符号结束，默认是逗号
--lines-terminated-by char	设定每行记录之间的分隔符，默认是\n
--mysql-delimiters	MySQL 默认的分隔符设置，字段之间以逗号分隔，行之间以\n 分隔，默认转义符是\，字段以单引号包裹
--optionally-enclosed-by char	给带有双引号或单引号的字段前后加上指定字符

10.3.4 Sqoop import 命令的基本操作

（1）创建 MySQL 数据库 test_sqoop。

进入 MySQL。

```
[root@qf01 ~]# mysql -uroot -p
```

创建数据库 test_sqoop。

```
mysql> create database test_sqoop;
```

（2）在数据库 test_sqoop 中创建表 sqoop_table。

使用数据库 test_sqoop。

```
mysql> use test_sqoop;
```

创建表 sqoop_table。

```
mysql> create table sqoop_table(
    -> id int,
    -> name varchar(40),
    -> age int,
    -> primary key(id));
```

向表 sqoop_table 中插入数据。

```
mysql> insert into sqoop_table
    -> (id, name, age)
    -> values
    -> (1, "Jack", 22);
```

（3）在虚拟机 qf01 中，使用 Sqoop 命令列出 MySQL 的数据库。

```
[root@qf01 ~]# sqoop list-databases \
--connect jdbc:mysql://qf01:3306 --username root -P
18/10/29 13:04:09 INFO sqoop.Sqoop: Running Sqoop version: 1.4.7
Enter password:
18/10/29 13:04:12 INFO manager.MySQLManager: Preparing to use a MySQL streaming resultset.
information_schema
hive
mysql
performance_schema
test_sqoop
```

提示：命令中的 username 表示 MySQL 的用户名，-P 表示 MySQL 的密码。

（4）列出 MySQL 中的数据库 test_sqoop 下的所有表。

```
[root@qf01 ~]# sqoop list-tables \
--connect jdbc:mysql:// qf01:3306/test_sqoop --username root -P
18/10/29 13:45:16 INFO sqoop.Sqoop: Running Sqoop version: 1.4.7
Enter password:
18/10/29 13:45:19 INFO manager.MySQLManager: Preparing to use a MySQL streaming resultset.
sqoop_table
```

10.4 Sqoop 数据导入

10.4.1 将 MySQL 的数据导入 HDFS

将 MySQL 的数据导入 HDFS，操作如下。

(1)确保 ZooKeeper 和 Hadoop 集群已启动。
(2)在 MySQL 中给虚拟机 qf01、qf02 和 qf03 授予访问权限。

```
mysql> grant all PRIVILEGES on *.* to root@'qf01' identified by 'root';
mysql> grant all PRIVILEGES on *.* to root@'qf02' identified by 'root';
mysql> grant all PRIVILEGES on *.* to root@'qf03' identified by 'root';
```

(3)将 MySQL 数据库 test_sqoop 下的表 sqoop_table 中的数据导入 HDFS。

```
[root@qf01 ~]# sqoop import \
--connect jdbc:mysql://qf01:3306/test_sqoop --username root -P \
--table sqoop_table \
--target-dir /sqoop/sqoop_table -m 2
```

① 将 MySQL 的数据导入 HDFS 后,可以在当前目录下查看 Sqoop 自动生成的 Java 程序(sqoop_table.java)。

```
[root@qf01 ~]# cat sqoop_table.java
```

② 查看导入 HDFS 的/sqoop/sqoop_table 目录下的数据。

```
[root@qf01 ~]# hdfs dfs -cat /sqoop/sqoop_table/*
1,Jack,22
```

出现以上数据表明通过 Sqoop 命令将 MySQL 的数据成功导入了 HDFS。
(4)将 MySQL 的数据导入 HDFS,并指定字段分隔符。

```
[root@qf01 ~]# sqoop import \
--connect jdbc:mysql://qf01:3306/test_sqoop --username root -P \
--table sqoop_table --target-dir /sqoop/sqoop_table -m 2 \
--fields-terminated-by '\t' --delete-target-dir
```

查看导入 HDFS 的/sqoop/sqoop_table 目录下的数据。

```
[root@qf01 ~]# hdfs dfs -cat /sqoop/sqoop_table/*
1    Jack    22
```

(5)将 MySQL 的表 sqoop_table 的指定列导入 HDFS。

```
[root@qf01 ~]# sqoop import \
--connect jdbc:mysql:// qf01:3306/test_sqoop --username root -P \
--table sqoop_table --target-dir /sqoop/sqoop_table -m 2 \
--fields-terminated-by '\t' --delete-target-dir \
--columns id,name
```

查看导入 HDFS 的/sqoop/sqoop_table 目录下的数据。

```
[root@qf01 ~]# hdfs dfs -cat /sqoop/sqoop_table/*
1    Jack
```

10.4.2 将 MySQL 的数据导入 Hive

Sqoop import 命令中与 Hive 相关的常用参数,如表 10.5 所示。

表 10.5　　Sqoop import 命令中与 Hive 相关的常用参数

参数	说明
--create-hive-table	创建 Hive 表
--external-table-dir <hdfs path>	外部表路径
--hive-database <database-name>	Hive 数据库
--hive-delims-replacement <arg>	行分隔符
--hive-import	导入 Hive 表指定的参数
--hive-overwrite	替换 Hive 中已存在的表
--hive-partition-key <partition-key>	指定 Hive 分区字段
--hive-partition-value <partition-value>	指定 Hive 分区值
--hive-table <table-name>	指定 Hive 导入表
--map-column-hive <arg>	指定 Hive 列映射

将 MySQL 的数据导入 Hive 的具体操作如下。

(1) 修改相关配置。

① 将/mysoft/hive/conf 目录下的 Hive 配置文件 hive-site.xml 复制到/mysoft/sqoop/conf 目录下。

```
[root@qf01 ~]# cp /mysoft/hive/conf/hive-site.xml /mysoft/sqoop/conf/
```

② 修改/etc/profile 文件。

```
[root@qf01 ~]# vi /etc/profile
```

添加如下内容。

```
export HADOOP_CLASSPATH=$HADOOP_CLASSPATH:$HIVE_HOME/lib/*
```

使/etc/profile 文件生效。

```
[root@qf01 ~]# source /etc/profile
```

③ 修改/usr/java/jdk1.8.0_121/jre/lib/security/java.policy 文件。

```
[root@qf01 ~]# vi /usr/java/jdk1.8.0_121/jre/lib/security/java.policy
```

在最后一行 (};) 之前添加如下内容。

```
permission javax.management.MBeanTrustPermission "register";
```

(2) 在虚拟机 qf01 上，启动 Hiveserver2 服务。

```
[root@qf01 ~]# hiveserver2
...
SLF4J: Actual binding is of type [org.apache.logging.slf4j.Log4jLoggerFactory]
```

(3) 在虚拟机 qf01 上，重新打开一个终端，启动 Beeline 客户端。

```
[root@qf01 ~]# beeline -u jdbc:hive2://localhost:10000 -n root
...
0: jdbc:hive2://localhost:10000>
```

(4) 在虚拟机 qf01 上，再重新打开一个终端，根据 MySQL 数据库 test_sqoop 下表 sqoop_table

的表结构，创建 Hive 表 sqoop_hive_table。

```
[root@qf01 ~]# sqoop create-hive-table \
--connect jdbc:mysql://qf01:3306/test_sqoop --username root -P \
--table sqoop_table --hive-table sqoop_hive_table
...
18/10/31 17:40:32 INFO hive.HiveImport: Hive import complete.
```

在 Beeline 客户端查看表 sqoop_hive_table 的表结构。

```
jdbc:hive2://> desc sqoop_hive_table;
+-----------+------------+----------+--+
| col_name  | data_type  | comment  |
+-----------+------------+----------+--+
| id        | int        |          |
| name      | string     |          |
| age       | int        |          |
+-----------+------------+----------+--+
3 rows selected (0.265 seconds)
```

出现以上数据表明成功创建了 Hive 表 sqoop_hive_table。

（5）创建 Hive 分区表。

注意：此处的 Hive 分区表只能创建一个分区。

```
[root@qf01 ~]# sqoop create-hive-table \
--connect jdbc:mysql://qf01:3306/test_sqoop --username root -P \
--table sqoop_table --hive-database test --hive-table sqoop_hive_table2 \
--fields-terminated-by '\t' \
--hive-partition-key city
```

在 Beeline 客户端查看表 sqoop_hive_table2 的表结构。

① 使用 Hive 数据库 test。

```
jdbc:hive2://> use test;
No rows affected (2.035 seconds)
```

② 查看表 sqoop_hive_table2 的表结构。

```
jdbc:hive2://> desc sqoop_hive_table2;
+--------------------------+-----------------------+-----------------------+--+
|         col_name         |       data_type       |        comment        |
+--------------------------+-----------------------+-----------------------+--+
| id                       | int                   |                       |
| name                     | string                |                       |
| age                      | int                   |                       |
| city                     | string                |                       |
|                          | NULL                  | NULL                  |
| # Partition Information  | NULL                  | NULL                  |
| # col_name               | data_type             | comment               |
|                          | NULL                  | NULL                  |
| city                     | string                |                       |
+--------------------------+-----------------------+-----------------------+--+
9 rows selected (1.302 seconds)
```

出现以上数据表明 Sqoop 成功地创建了 Hive 分区表 sqoop_hive_table2，并且添加了分区字段 city（默认为字符串类型）。

（6）将 MySQL 数据库 test_sqoop 下表 sqoop_table 中的数据导入 Hive 表 sqoop_hive_table3。

```
[root@qf01 ~]# sqoop import \
--connect jdbc:mysql://qf01:3306/test_sqoop --username root -P \
--table sqoop_table --hive-import --hive-table sqoop_hive_table3 \
--fields-terminated-by '\t'
```

在 Beeline 客户端查看表 sqoop_hive_table3 的数据。

① 使用 Hive 默认数据库。

```
jdbc:hive2://> use default;
No rows affected (0.111 seconds)
```

② 查看表 sqoop_hive_table3 的数据。

```
jdbc:hive2://> select * from sqoop_hive_table3;
+----------------------+------------------------+-----------------------+--+
| sqoop_hive_table3.id | sqoop_hive_table3.name | sqoop_hive_table3.age |
+----------------------+------------------------+-----------------------+--+
| 1                    | Jack                   | 22                    |
+----------------------+------------------------+-----------------------+--+
1 row selected (0.288 seconds)
```

出现以上数据表明 Sqoop 成功地将 MySQL 的数据导入了指定的 Hive 表。

（7）将 MySQL 的表 sqoop_table 中的数据导入 Hive 的分区表 sqoop_hive_table4。注意：MySQL 的数据导入 Hive 分区表时，只能指定一个分区。

```
[root@qf01 ~]# sqoop import \
--connect jdbc:mysql://qf01:3306/test_sqoop --username root -P \
--table sqoop_table --hive-import --hive-table sqoop_hive_table4 \
--fields-terminated-by '\t' \
--hive-partition-key city --hive-partition-value Beijing \
--hive-database test
```

在 Beeline 客户端查看表 sqoop_hive_table4 的数据。

① 使用 Hive 默认数据库。

```
jdbc:hive2://> use test;
No rows affected (0.111 seconds)
```

② 查看表 sqoop_hive_table4 的数据。

```
jdbc:hive2://> select * from sqoop_hive_table4;
+----------------------+------------------------+-----------------------+---------
----------------+--+
| sqoop_hive_table4.id | sqoop_hive_table4.name | sqoop_hive_table4.age | sqoop_hive_table4.city |
+----------------------+------------------------+-----------------------+---------
----------------+--+
| 1                    | Jack                   | 22                    | Beijing|
+----------------------+------------------------+-----------------------+---------
----------------+--+
1 row selected (1.37 seconds)
```

出现以上数据表明 Sqoop 成功地将 MySQL 的数据导入了指定的 Hive 分区表。

10.4.3　将 MySQL 的数据导入 HBase

Sqoop import 命令中与 HBase 相关的常用参数，如表 10.6 所示。

表 10.6　　Sqoop import 命令中与 HBase 相关的常用参数

参数	说明
--column-family <family>	为导入的数据设置目标列簇
--hbase-bulkload	启用 HBase 批量加载
--hbase-create-table	创建未存在的目标 HBase 表
--hbase-row-key <col>	指定用作 Rowkey 的输入列
--hbase-table <table>	指定导入数据的目标 HBase 表

将 MySQL 的数据导入 HBase 的具体操作如下。

（1）关闭 Hiveserver2 和 Beeline 客户端，启动 HBase。

```
[root@qf01 ~]# start-hbase.sh
```

（2）将 MySQL 数据库 test_sqoop 下表 sqoop_table 的数据导入 HBase 表 sqoop_hbase_table 中。

```
[root@qf01 ~]# sqoop import \
> --connect jdbc:mysql://qf01:3306/test_sqoop --username root -P \
> --table sqoop_table --hbase-create-table --hbase-table sqoop_hbase_table \
> --hbase-row-key id --column-family f1
```

（3）启动 HBase Shell 命令行，查看 HBase 表 sqoop_hbase_table 的数据。

```
[root@qf01 ~]# hbase shell
hbase(main)> scan 'sqoop_hbase_table'
ROW                  COLUMN+CELL
 1                   column=f1:age, timestamp=1541044186956, value=22
 1                   column=f1:name, timestamp=1541044186956, value=Jack
1 row(s) in 0.2640 seconds
```

出现以上数据表明 Sqoop 成功地将 MySQL 的数据导入了指定的 HBase 表。

10.4.4　增量导入

Sqoop 命令中与增量导入相关的参数，如表 10.7 所示。

表 10.7　　Sqoop 命令中与增量导入相关的参数

参数	说明
--check-column <column>	用来指定一些列，在进行增量导入时，导入这些列的数据
--incremental <import-type>	用来指定增量导入的模式，有两种模式：append 和 lastmodified。 （1）当表以 id 连续增加导入新的记录时，使用 append 模式。 （2）当表执行更新操作时，使用 lastmodified 模式
--last-value <value>	指定上一次导入操作后指定列的最大值

（1）将 MySQL 的数据增量导入 HDFS。

① 在 MySQL 数据库 test_sqoop 下的表 sqoop_table 中插入 3 条新的数据。

```
mysql> use test_sqoop;
mysql> insert into sqoop_table
    -> (id, name, age)
    -> values
    -> (2, "Sophie", 23),
```

```
    -> (3, "Tom", 24),
    -> (4, "Helen", 25);
Query OK, 3 rows affected (0.01 sec)
Records: 3  Duplicates: 0  Warnings: 0
```

② 将 MySQL 数据库 test_sqoop 下的表 sqoop_table 中的数据增量导入 HDFS 的/sqoop/sqoop_table 目录下。

```
[root@qf01 ~]# sqoop import \
> --connect jdbc:mysql://qf01:3306/test_sqoop --username root -P \
> --table sqoop_table \
> --check-column id --incremental append --last-value 1 \
> --target-dir /sqoop/sqoop_table -m 2
> -fields-terminated-by '\t'
```

③ 查看导入 HDFS 的/sqoop/sqoop_table 目录下的数据。

```
[root@qf01 ~]# hdfs dfs -cat /sqoop/sqoop_table/*
1       jack
2       Sophie  23
3       Tom     24
4       Helen   25
```

提示：第 1 行是先前导入的数据。

出现以上数据表明通过 Sqoop 命令将 MySQL 的数据成功增量导入了 HDFS。

（2）将 MySQL 的数据增量导入 Hive。将 MySQL 的数据增量导入 Hive 时，默认实现的是将 MySQL 指定表的全部数据增量导入 Hive 指定表，然而实际工作中需要的是只把 MySQL 指定表增加的数据增量导入 Hive 指定表，下面对此种情形进行演示。

① 启动 Hiveserver2 服务和 Beeline 客户端（需要重新打开一个终端）。

```
[root@qf01 ~]# hiveserver2
```

② 将 MySQL 数据库 test_sqoop 下表 sqoop_table 中的数据增量导入 Hive 表 sqoop_hive_table3。

```
[root@qf01 ~]# sqoop import \
> --connect jdbc:mysql://qf01:3306/test_sqoop --username root -P \
> --table sqoop_table --hive-import --hive-table sqoop_hive_table3 \
> --fields-terminated-by '\t' \
> --check-column id --incremental append --last-value 1
```

提示：通过参数--check-column id --incremental append --last-value 1 指定将 MySQL 表 sqoop_table 中 id 为 1 的一行数据之后的数据增量导入 Hive 表 sqoop_hive_table3。

③ 在 Beeline 客户端查看表 sqoop_hive_table3 的数据。

使用 Hive 默认数据库。

```
jdbc:hive2://> use default;
No rows affected (0.111 seconds)
```

查看表 sqoop_hive_table3 的数据。

```
jdbc:hive2://> select * from sqoop_hive_table3;
+----------------------+------------------------+-----------------------+--+
| sqoop_hive_table3.id | sqoop_hive_table3.name | sqoop_hive_table3.age |
+----------------------+------------------------+-----------------------+--+
| 1                    | jack                   | 22                    |
```

```
| 2                     | Sophie                    | 23                       |   |
| 3                     | Tom                       | 24                       |   |
| 4                     | Helen                     | 25                       |   |
+-----------------------+---------------------------+--------------------------+---+
4 rows selected (3.078 seconds)
```

出现以上数据表明 Sqoop 成功地将 MySQL 的数据增量导入了指定的 Hive 表。

（3）将 MySQL 的数据增量导入 HBase。

① 关闭 Hiveserver2 和 Beeline 客户端，启动 HBase。

```
[root@qf01 ~]# start-hbase.sh
```

② 将 MySQL 数据库 test_sqoop 下表 sqoop_table 的数据增量导入 HBase 表 sqoop_hbase_table 中。

```
[root@qf01 ~]# sqoop import \
> --connect jdbc:mysql://qf01:3306/test_sqoop --username root -P \
> --table sqoop_table --hbase-create-table --hbase-table sqoop_hbase_table \
> --hbase-row-key id --column-family f1 \
> --check-column id --incremental append --last-value 1
```

③ 启动 HBase Shell 命令行，查看 HBase 表 sqoop_hbase_table 的数据。

```
[root@qf01 ~]# hbase shell
hbase(main)> scan 'sqoop_hbase_table'
ROW                    COLUMN+CELL
 1                      column=f1:age, timestamp=1541044186956, value=22
 1                      column=f1:name, timestamp=1541044186956, value=jack
 2                      column=f1:age, timestamp=1541128279046, value=23
 2                      column=f1:name, timestamp=1541128279046, value=Sophie
 3                      column=f1:age, timestamp=1541128288905, value=24
 3                      column=f1:name, timestamp=1541128288905, value=Tom
 4                      column=f1:age, timestamp=1541128285868, value=25
 4                      column=f1:name, timestamp=1541128285868, value=Helen
4 row(s) in 1.0470 seconds
```

出现以上数据表明 Sqoop 成功地将 MySQL 的数据增量导入了指定的 HBase 表。

10.4.5 按需导入

将 MySQL 的数据按需导入 HDFS/Hive/HBase，使用的参数是--query，该参数主要用于通过查询语句得到需要导入目标表的数据。

将 MySQL 数据库 test_sqoop 下的表 sqoop_table 中 id 大于 2 的数据导入 HDFS 的 /sqoop/sqoop_query_table 目录下。

```
[root@qf01 ~]# sqoop import \
> --connect jdbc:mysql://qf01:3306/test_sqoop --username root -P \
> --query 'select * from sqoop_table where id>2 and $CONDITIONS' \
> --split-by id --fields-terminated-by '\t' -m 2 \
> --target-dir /sqoop/sqoop_query_table
```

查看导入 HDFS 的/sqoop/sqoop_query_table 目录下的数据。

```
[root@qf01 ~]# hdfs dfs -cat /sqoop/sqoop_query_table/*
...
3       Tom     24
4       Helen   25
```

出现以上数据表明 Sqoop 成功地将 MySQL 的数据按需导入了 HDFS。

10.5 Sqoop 数据导出

10.5.1 将 HDFS 的数据导出到 MySQL

以将 HDFS 的数据导出到 MySQL 为例，讲解 Sqoop export 命令。将 HDFS 的数据导出到 MySQL 前，必须先确认目标表存在于 MySQL 中，如果目标表未存在，需要先创建表结构。

HDFS 的数据导出到 MySQL 的默认操作是生成 insert 语句将数据插入指定的 MySQL 表；在更新模式下，则生成 update 语句更新指定的 MySQL 表的数据。

将 HDFS 的数据导出到 MySQL，具体操作步骤如下。

（1）准备数据。

① 新建本地文件 sqoopdata.txt。

```
[root@qf01 ~]# vi sqoopdata.txt
```

添加以下内容。**提示**：数据之间的分隔符为制表符。

```
1101  Sophie    30   Shanghai  China
1106  Helen     35   Paris     France
1108  Tom  27   Ottawa    Canada
1109  Jack 25   London    Britain
1103  Carlos    26   Washington   American
```

② 把本地文件 sqoopdata.txt 上传到 HDFS 的/sqoop 目录下。

```
[root@qf01 ~]# hdfs dfs -put sqoopdata.txt /sqoop
```

（2）创建 MySQL 中的目标表。

① 在虚拟机 qf01 上，重新打开一个终端，启动 MySQL。

```
[root@qf01 ~]# mysql -uroot -p
Enter password:
...
mysql>
```

② 在 MySQL 数据库 test_sqoop 下创建目标表 student。

```
mysql> use test_sqoop;
mysql> create table student (
    -> id int,
    -> name varchar(20),
    -> age int,
    -> city varchar(30),
    -> country varchar(30));
Query OK, 0 rows affected (0.04 sec)
```

（3）将 HDFS 的/sqoop/sqoopdata.txt 文件中的数据导出到 MySQL 数据库 test_sqoop 下的表 student 中。

```
[root@qf01 ~]# sqoop export \
> --connect jdbc:mysql://qf01:3306/test_sqoop --username root -P \
```

```
> --table student --export-dir /sqoop/sqoopdata.txt \
> --fields-terminated-by '\t'
```

（4）查看 MySQL 数据库 test_sqoop 下表 student 的数据。

```
 mysql> use test_sqoop;
mysql> select * from student;
+------+--------+------+------------+----------+
| id   | name   | age  | city       | country  |
+------+--------+------+------------+----------+
| 1101 | Sophie |  30  | Shanghai   | China    |
| 1108 | Tom    |  27  | Ottawa     | Canada   |
| 1109 | Jack   |  25  | London     | Britain  |
| 1106 | Helen  |  35  | Paris      | France   |
| 1103 | Carlos |  26  | Washington | American |
+------+--------+------+------------+----------+
5 rows in set (0.01 sec)
```

出现以上数据表明 Sqoop 成功地将 HDFS 的数据导出到了 MySQL。

10.5.2 将 Hive 的数据导出到 MySQL

（1）将 Hive 的数据导出到 MySQL。

```
[root@qf01 ~]# sqoop export \
> --connect jdbc:mysql://qf01:3306/test_sqoop --username root -P \
> --table student --export-dir /user/hive/warehouse/sqoop_hive_table3 \
> --input-fields-terminated-by '\t'
```

（2）查看 MySQL 数据库 test_sqoop 下表 student 的数据。

```
mysql> use test_sqoop;
mysql> mysql> select * from student;
+------+--------+------+------------+----------+
| id   | name   | age  | city       | country  |
+------+--------+------+------------+----------+
| 1101 | Sophie |  30  | Shanghai   | China    |
| 1108 | Tom    |  27  | Ottawa     | Canada   |
| 1109 | Jack   |  25  | London     | Britain  |
| 1106 | Helen  |  35  | Paris      | France   |
| 1103 | carlos |  26  | Washington | American |
|    1 | jack   |  22  | NULL       | NULL     |
+------+--------+------+------------+----------+
6 rows in set (0.02 sec)
```

出现以上数据表明 Sqoop 成功地将 Hive 的数据导出到了 MySQL。

10.5.3 将 HBase 的数据导出到 MySQL

将 HBase 的数据导出到 MySQL，需要借助 Hive 的中转作用来完成，数据传输的一般路径是 HBase→Hive 外部表→Hive 内部表→MySQL，具体的操作步骤如下。

（1）在 MySQL 数据库 test_sqoop 下创建目标表 user。

```
mysql> use test_sqoop;
mysql> create table user (
    -> rowkey int,
    -> id int,
    -> name varchar(20),
```

```
    -> primary key (id));
Query OK, 0 rows affected (0.04 sec)
```

（2）创建 HBase 表。

① 启动 HBase 和 HBase Shell 客户端。

```
[root@qf01 ~]# start-hbase.sh
[root@qf01 ~]# hbase shell
...
hbase(main)>
```

② 创建 HBase 表 user。

```
hbase(main)> create 'user', 'data'
hbase(main)> put 'user', 1, 'data:id', 1
hbase(main)> put 'user', 1, 'data:name', 'Jack'
hbase(main)> put 'user', 2, 'data:id', 2
hbase(main)> put 'user', 2, 'data:name', 'Tom'
```

（3）创建 Hive 外部表。

① 在虚拟机 qf01 上，启动 Hiveserver2 服务。

```
[root@qf01 ~]# hiveserver2
```

② 在虚拟机 qf01 上，重新打开一个终端，启动 Beeline 客户端。

```
[root@qf01 ~]# beeline -u jdbc:hive2://localhost:10000 -n root
```

③ 创建 Hive 外部表 sqoop_hive_user。

```
jdbc:hive2://> create external table test.sqoop_hive_user(key int, id int, name string)
stored by 'org.apache.hadoop.hive.hbase.HBaseStorageHandler'
with serdeproperties("hbase.columns.mapping" = ":key, data:id, data:name")
tblproperties("hbase.table.name" = "user",
"hbase.mapred.output.outputtable" = "user");
```

（4）创建 Hive 内部表。

```
jdbc:hive2://> create table hive_user(key int,id int,name string);
```

（5）将 Hive 外部表的数据导入内部表。

```
hive> insert overwrite table hive_user select * from test.sqoop_hive_user;
```

（6）将 Hive 内部表的数据导出到 MySQL 数据库 test_sqoop 下表 user 中。

```
[root@qf01 ~]# sqoop export \
> --connect jdbc:mysql://qf01:3306/test_sqoop --username root -P \
> --table user --export-dir /user/hive/warehouse/hive_user \
> --input-fields-terminated-by '\001'
```

提示：\001 为 Hive 默认的字段分隔符。

查看 MySQL 数据库 test_sqoop 下表 user 的数据。

```
mysql> select * from user;
+--------+----+------+
| rowkey | id | name |
+--------+----+------+
```

```
|        1 |  1 | Jack  |
|        2 |  2 | Tom   |
+----------+----+-------+
2 rows in set (0.02 sec)
```

出现以上数据表明 Sqoop 成功地将 HBase 的数据导出到了 MySQL。

10.6　Sqoop job

Sqoop 的 job 命令主要用于创建和维护 Sqoop 作业（Job），以便保存 Sqoop 的导入和导出命令。

（1）创建 Sqoop 作业。创建 Sqoop 作业用于保存将 MySQL 的数据导入 HDFS 的命令。

```
[root@qf01 ~]# sqoop job --create myjob -- import \
> --connect jdbc:mysql://qf01:3306/test_sqoop --username root -P \
> --table sqoop_table --target-dir /sqoop/job/sqoop_table
```

（2）列出现有的 Sqoop 作业。

```
[root@qf01 ~]# sqoop job --list
18/11/05 00:16:07 INFO sqoop.Sqoop: Running Sqoop version: 1.4.7
Available jobs:
myjob
```

（3）执行 Sqoop 作业。执行 Sqoop 作业 myjob，即执行先前保存的将 MySQL 的数据导入 HDFS 的命令。

```
[root@qf01 ~]# sqoop job --exec myjob
```

查看 Sqoop 作业执行的结果。

```
[root@qf01 ~]# hdfs dfs -cat /sqoop/job/sqoop_table/*
...
1,jack,22
2,Sophie,23
3,Tom,24
4,Helen,25
```

出现以上数据表明 Sqoop 成功地执行了 myjob 作业，即成功地执行了将 MySQL 的数据导入 HDFS 的命令。

（4）删除 Sqoop 作业。

```
[root@qf01 ~]# sqoop job --delete myjob
```

10.7　本章小结

本章主要对 Sqoop 的原理、架构、安装、命令进行了讲解，核心内容是 import 和 export 命令，应重点掌握其用法。使用 Sqoop 命令，可以方便地实现数据在关系型数据库和 Hadoop 组件（HDFS/Hive/HBase）之间的传输，这些命令大大提高了大数据开发工作的效率。

10.8 习题

1. 填空题

（1）_____是一种用于在 Hadoop 和结构化数据系统（如关系型数据库、大型机）之间高效传输数据的工具。

（2）Sqoop 项目开始于_____年。

（3）Sqoop 的导入和导出功能就是通过_____作业来实现的。

（4）目前 Sqoop 主要有两个系列，分别为_____和_____。

2. 选择题

（1）下面属于 Sqoop 数据库连接参数的有（　　）。

    A. --connect   B. --P   C. --help   D. --username

（2）列出 MySQL 数据库的 Sqoop 命令是（　　）。

    A. sqoop list-databases   B. sqoop list_table

    C. sqoop list   D. sqoop list_command

（3）Sqoop import 命令中与 Hive 相关的常用参数有（　　）。

    A. --hive-import   B. --create-hive-table

    C. --hive-table   D. --create-hive-list

（4）Sqoop import 命令中与 HBase 相关的常用参数有（　　）。

    A. --column-family   B. --hbase-create-table

    C. --hbase-row-key   D. --hbase-table

（5）将 MySQL 的数据按需导入 HDFS/Hive/HBase，使用的参数是（　　）。

    A. --where   B. --need   C. --query   D. --alter

3. 思考题

（1）简述 Sqoop 导入导出数据操作流程。

（2）将 HBase 的数据导出到 MySQL，数据传输的一般路径是什么？

第11章 综合项目——电商精准营销

本章学习目标
- 了解项目背景及需求
- 了解项目中的架构设计
- 了解数据来源
- 掌握数据清洗流程
- 掌握数据仓库操作流程
- 掌握应用测试方法

在前面的章节中，我们详细地讲解了 Hadoop 的基础知识与核心技术，以及各相关组件的使用，包括 Hadoop、Hive、HBase、ZooKeeper、Flume、Sqoop 等。只有把理论知识同具体实际相结合，才能正确回答实践提出的问题，扎实提升读者的理论水平与实战能力。本章通过一个企业级真实项目案例，串联前面学习的知识点，讲解这些知识点在实际开发过程中的应用。通过本章的学习，读者可以真正理解 Hadoop 的精髓，并做到融会贯通、学以致用。

11.1 项目概述

11.1.1 项目背景介绍

电商网站上线之后，利用大数据技术，收集用户的行为数据，进行多维度统计分析，掌握网站线上运营情况，将分析结果生成相应的数据报表，提供给运营部门进行业务分析。运营部门利用数据报表，可以制订出相应的网站优化方案，调整广告投入，组织举办更好的促销、精准营销等活动。

11.1.2 项目架构设计

项目架构：数据源（JS、SDK）——数据采集（Flume）——数据预处理（MapReduce）——数据仓库（Hive）——数据导出（Sqoop）——数据存储（MySQL）——数据可视化。

数据分析平台结构如图 11.1 所示。

（1）数据采集。每当用户通过 PC 端或者移动端访问电商网站时，网站前台后台程序均会产生日志信息，前台信息通过 JS(JavaScript)收集到 Nginx

服务器中,后台信息通过 SDK(Software Development Kit,软件开发工具包)收集到 Nginx 服务器中。然后在 Nginx 服务器中部署 Flume Agent 采集软件,实时监控目录,将产生的日志文件实时采集到 HDFS 当中。

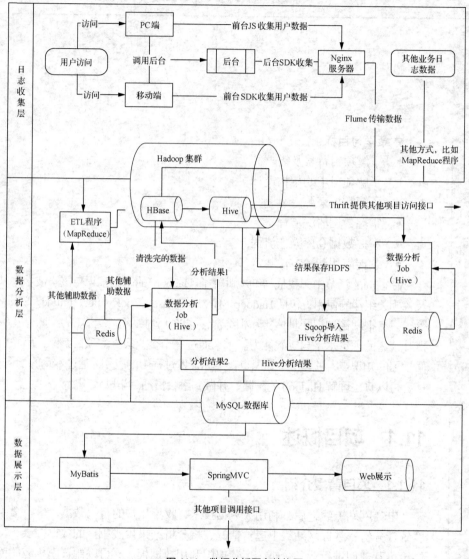

图 11.1 数据分析平台结构图

(2)数据预处理。将采集到的数据按照项目需求进行初步的清洗,得到项目中需要使用的字段数据。

(3)数据仓库。由于 MapReduce 操作数据编写流程过于复杂,这里采用 Hive 来对数据进行处理。将数据导入 Hive,按照项目的需求,写 SQL 语句来实现。

(4)数据存储。处理之前的数据以及 Hive 的输出数据都存储在 HDFS 中,读取十分缓慢,很容易造成超时,这里采用 Sqoop 工具,将数据导入 MySQL。

(5)数据可视化。为了更直观地展示数据结果,对得到的数据结果进行可视化操作。

11.2 项目详细介绍

11.2.1 项目核心关注点

本项目收集不同客户端的用户行为数据，通过 MapReduce、Hive 进行数据分析处理，将分析结果数据保存到关系型数据库中。在此过程中需要对几个核心关注点进行详细分析。

（1）购买率：购买的人数/总人数，购买的人数/查看该商品的总人数。
（2）复购率：n 次购买的人数 / $n-1$ 次购买的人数（$n \geq 2$）。
（3）订单数量，订单金额，订单的类型。
① 成功订单数量，成功订单金额，成功订单的类型。
② 退款订单数量，退款订单金额，退款订单的类型。
（4）访客人数/会员人数。
（5）访客转会员的比例。
（6）广告推广效果。
（7）网站内容相关分析（网站的跳出率，页面的跳出率）。

11.2.2 重要概念

1. 访客

访问网站的用户，一般称为自然人，区分 PC 用户和手机用户。

PC 用户以 IP 地址标识。由于 NAT、代理等情况的存在，一个 IP 地址可能对应多个访客，但它实现简单，采用客户端种植 Cookie 的方式，当用户第一次访问系统时，在客户端的 Cookie 中保存一个 UUID（Universally Unique Identifier，通用唯一识别码），将过期时间设置为 10 年。

手机用户以设备的固定识别码标识，如 IMEI（International Mobile Equipment Identity，国际移动设备识别码）、MEID（Mobile Equipment Identifier，移动设备识别码）等。个别情况下，这些识别码也可能会出现误差，多个设备对应一个识别码。用户第一次访问系统的时候，磁盘中会写入标识符。

访客统计指标如下。
（1）新增访客：第一次访问系统的访客人数。
（2）活跃访客：给定时间段内访问过系统的访客人数（老访客+新访客）。
（3）总访客：迄今为止访问过系统的访客总人数。
（4）流失访客：上一个时间段内访问过系统，当前时间段内没有访问系统的访客人数。
（5）回流访客：上一个时间段内没有访问过系统，当前时间段内访问过系统的访客人数。

2. 会员

业务系统的注册用户，直接使用业务系统中的会员标识符来标识。访客登录系统后，就成为会员。

会员统计指标如下。
（1）新增会员：第一次登录系统的会员人数。
（2）活跃会员：给定时间段内登录过系统的会员人数（老会员+新会员）。
（3）总会员：迄今为止新增会员的总人数。
（4）回流会员：

（5）流失会员。
（6）访客转会员比例。
（7）新增访客转会员比例。

3. 会话

用户进入系统到离开系统这一段时间被称为会话，这段时间的长度就叫作会话长度。一个会话中的所有操作都属于这个会话。会话分为 PC 端会话和移动端会话。

PC 端会话采用浏览器的 Session 机制在 Cookie 中存储一个存活时间，在操作的时候，先判断会话是否过期，如果过期，产生一个新的会话，如果没有过期，更新存活时间。

移动端会话采用移动端的 Session 机制，类似 PC 端种植 Cookie 的方式，在磁盘中写入一个时间进行判断。

会话统计指标如下。
（1）会话长度。
（2）会话数量。
（3）跳出会话数量（在一个会话中，只访问过一次网站的会话数量）。

4. 跳出率

跳出率统计指标如下。
（1）会话跳出率：跳出会话数量/总会话数量。
（2）页面跳出率：从该页面离开后进入其他页面的会话数量占进入该页面会话数量的百分比。

5. 外链

外链统计指标：不同外链带来的会话数量/访客数量/订单数量。

6. PV

PV（Page View，页面访问量）是用户对页面的访问总次数。用户每次对网站的访问均被记录，用户多次访问同一页面，访问量累计。

7. UV

统计 UV（Unique Visitor，独立访问用户）数量。访问网站的一台计算机为一个访客，00:00—24:00 相同客户端只被计算一次。

8. 独立 IP

统计独立 IP 数量。00:00—24:00 相同 IP 地址只被计算一次。

9. DV

DV（Depth View，访问深度）指访问了多少页面，展示网站内容对用户的吸引程度，结合跳出率，有助于修改网站内容，提高网站黏性和友好性。统计指标：不同访问深度的访客人数/会话数量。

11.2.3 维度

数据分析只有基于相应的维度才有意义。常见的维度如表 11.1 所示。

表 11.1　　　　　　　　　　　　　常见维度

维度	解释
时间维度	年、季度、月、周、日、小时等
平台维度	PC 端、移动端、程序后台等

续表

维度	解释
浏览器维度	区分浏览器类型、浏览器版本
地域维度	国家、省份、城市等
KPI 维度	指定分析的指标
版本维度	如 V1、V2 等，一般用于多个版本之间的数据比较（AB 测试）
支付方式维度	Alipay、WeChat Pay、银行卡支付等
外链维度	百度、360、Google 等
操作系统维度	操作系统名称、操作系统版本等

11.3 项目模块分析

本项目主要有七个数据分析模块，分别为：用户基本信息分析模块、浏览器分析模块、地域分析模块、外链分析模块、用户浏览深度分析模块、事件分析模块、订单分析模块。

针对不同的分析模块，我们有不同的用户数据需求，七个模块中，用户基本信息分析模块和浏览器分析模块类型一致，只是后者比前者多一个浏览器维度。地域分析模块和外链分析模块分别从不同的维度进行分析展示。用户浏览深度分析模块、事件分析模块以及订单分析模块是单独针对业务进行的分析。

11.3.1 用户基本信息分析模块

主要从访客和会员两个角度分析浏览相关信息，包括但不限于新增访客、活跃访客、总访客、新增会员、活跃会员、总会员以及会话分析等。

11.3.2 浏览器分析模块

在用户基本信息分析的基础上添加一个浏览器维度。
（1）浏览器访客分析。
（2）浏览器会员分析。
（3）浏览器会话分析。
（4）浏览器 PV 分析。

11.3.3 地域分析模块

主要分析各个不同省份的访客和会员状况。
（1）活跃访客地域分析。
（2）不同地域的跳出率分析。

11.3.4 外链分析模块

主要分析不同外链带来的用户访问量数据，如图 11.2 所示。
（1）外链偏好分析：分析各个外链带来的活跃访客数量。
（2）外链会话（跳出率）分析。

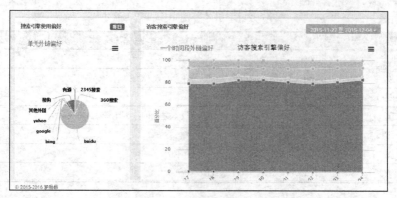

图 11.2　外链偏好分析图

11.4　数据采集

11.4.1　日志采集系统概述

我们需要将用户浏览行为的数据采集到我们的存储系统（HDFS）中，在本项目中，我们只收集 PC 端数据和程序后台的部分数据。在 PC 端，我们通过集成 JS 来收集用户浏览行为数据；在程序后台，我们通过集成 Java 的 jar 文件包来收集需要的数据。在这里只考虑 Java 开发环境。JS/jar 将收集的数据发送到 Nginx，然后 Flume 监控 Nginx 日志，将数据写入 HDFS，如图 11.3 所示。

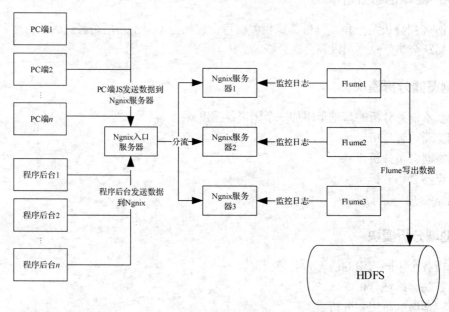

图 11.3　日志采集系统

11.4.2　JS SDK 收集数据

1. 概述

应了解 JS SDK 的集成方式以及提供的各种不同的 API。注意：不采用 IP 地址，而是通过在 Cookie 中填充一个 UUID 来标示用户的唯一性。

2. JS SDK 执行流程

JS SDK 中按照收集数据的不同有不同的事件,比如 Pageview 事件等。JS SDK 的执行流程如图 11.4 所示。

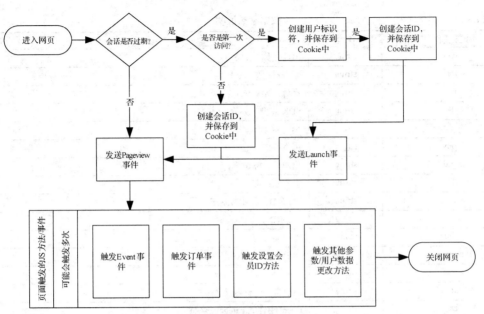

图 11.4　JS SDK 执行流程图

3. 程序前台事件分析

不同的分析模块需要不同的数据,下面分别分析每个模块需要的数据。用户基本信息就是用户的浏览行为信息,所以我们只需要 Pageview 事件。本项目涉及的事件如表 11.2 所示。

表 11.2　　　　　　　　　　　　　本项目涉及的事件

最终分析模块	PC 端 JS SDK 事件
用户基本信息分析	Pageview 事件
浏览器分析	
地域分析	
外链分析	
用户浏览深度分析	ChargeRequest 事件
订单分析	
事件分析	Event 事件
	Launch 事件

(1) Launch 事件。

用户第一次访问网站触发该事件。该事件不提供对外调用的接口,只实现该事件的数据收集。

(2) Pageview 事件。

用户访问页面/刷新页面触发该事件。该事件可以自动调用,也可以让程序员手动调用。

(3) ChargeRequest 事件。

用户下订单触发该事件。该事件需要程序主动调用。

(4) Event 事件。

用户触发业务定义的事件后,前端程序调用相关方法。

4. 数据参数说明

从各个不同事件中收集的不同数据被发送到 Nginx 服务器，但实际上这些数据还是有一些共性的。可能用到的数据参数说明如表 11.3 所示。

表 11.3 数据参数说明

参数名称	类型	描述
en	string	事件名称，如 e_pv
ver	string	版本号，如 0.0.1
pl	string	平台，如 website
sdk	string	SDK 类型，如 js
b_rst	string	浏览器分辨率
b_iev	string	浏览器信息
u_ud	string	用户唯一标识符
l	string	客户端语言
u_mid	string	会员 ID，和业务系统一致
u_sd	string	会话 ID
c_time	string	客户端时间
p_url	string	当前页面的 URL
p_ref	string	上一个页面的 URL
tt	string	当前页面的标题
ca	string	Event 事件的 category 名称
ac	string	Event 事件的 action 名称
kv_*	string	Event 事件的自定义属性
du	string	Event 事件的持续时间
oid	string	订单 ID
on	string	订单名称
cua	string	支付金额
cut	string	支付货币类型
pt	string	支付方式

11.4.3 Java SDK 收集数据

1. 概述

应了解 Java SDK 的集成方式以及提供的各种不同的方法。注意：由于在本次项目中 Java SDK 的作用主要就是发送支付成功/退款成功的信息给 Nginx 服务器，所以我们讲解的是一个简单版本的 Java SDK。

2. Java SDK 执行流程

Java SDK 支付成功执行流程如图 11.5 所示（退款成功与之类似）。

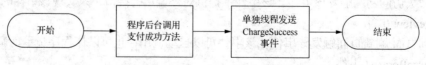

图 11.5 Java SDK 执行流程图

3. 程序后台事件分析

本项目中在程序后台只会触发 ChargeSuccess 事件，本事件的主要作用是发送支付成功的信息给

Nginx 服务器。发送方式与 PC 端发送方式相同，也是访问同一个 URL 来进行数据的传输。

（1）ChargeSuccess 事件。

会员最终支付成功触发该事件。该事件需要程序主动调用。

（2）ChargeRefund 事件。

会员进行退款操作触发该事件。该事件需要程序主动调用。

4. 数据参数说明

可能用到的数据参数说明如表 11.4 所示。

表 11.4 数据参数说明

参数名称	类型	描述
en	string	事件名称，如 e_cs
ver	string	版本号，如 0.0.1
pl	string	平台，如 website
sdk	string	SDK 类型，如 java
u_mid	string	会员 ID，和业务系统一致
c_time	string	客户端时间
oid	string	订单 ID

11.4.4 使用 Flume 搭建日志采集系统

由于 Flume 是部署在 Nginx 中采集日志数据的，所以在配置 Flume 之前，需要先对 Nginx 服务器进行配置，具体配置流程见附录。

下面开始配置 Flume，搭建日志采集系统。

（1）编写配置文件。在 Flume 的安装目录下的 conf 目录中，创建 flume-app.conf 文件，编辑内容如下。

```
#定义三大组件的名称
agent1.sources = source1
agent1.sinks = sink1
agent1.channels = channel1
# 配置 Source 组件
agent1.sources.source1.type = spooldir
agent1.sources.source1.spoolDir = /home/hadoop/logs/
agent1.sources.source1.fileHeader = false
#配置拦截器
agent1.sources.source1.interceptors = i1
agent1.sources.source1.interceptors.i1.type = host
agent1.sources.source1.interceptors.i1.hostHeader = hostname
# 配置 Sink 组件
agent1.sinks.sink1.type = hdfs
agent1.sinks.sink1.hdfs.path =hdfs://qianfeng01:9000/weblog/flume-collection/%y-%m-%d/%H-%M
agent1.sinks.sink1.hdfs.filePrefix = access_log
agent1.sinks.sink1.hdfs.maxOpenFiles = 5000
agent1.sinks.sink1.hdfs.batchSize= 100
agent1.sinks.sink1.hdfs.fileType = DataStream
agent1.sinks.sink1.hdfs.writeFormat =Text
agent1.sinks.sink1.hdfs.rollSize = 102400
```

```
agent1.sinks.sink1.hdfs.rollCount = 1000000
agent1.sinks.sink1.hdfs.rollInterval = 60
#agent1.sinks.sink1.hdfs.round = true
#agent1.sinks.sink1.hdfs.roundValue = 10
#agent1.sinks.sink1.hdfs.roundUnit = minute
agent1.sinks.sink1.hdfs.useLocalTimeStamp = true
# Use a channel which buffers events in memory
agent1.channels.channel1.type = memory
agent1.channels.channel1.keep-alive = 120
agent1.channels.channel1.capacity = 500000
agent1.channels.channel1.transactionCapacity = 600
# Bind the source and sink to the channel
agent1.sources.source1.channels = channel1
agent1.sinks.sink1.channel = channel1
```

依据官方文档进行配置，具体方法在本书第 9 章中有详细讲解。由于需要在 HDFS 中依据日期生成目录，存储抽取出来的数据，因此需要用**%y-%m-%d** 表达式进行表示。

（2）在 Flume 的安装目录下创建相应的文件夹。

```
mkdir spo_logs
mkdir filechannel
mkdir checkpoint
mkdir data
```

（3）运行 Flume。

```
bin/flume-ng agent \
--conf conf \
--name agent1 \
--conf-file conf/flume-app.conf \
-Dflume.root.logger=DEBUG,console
```

复制几份文件到 spo_logs 文件夹下，系统立即监测到该目录发生变化，实时抽取数据，而且.log 文件不会被抽取。至此 Flume 日志采集系统搭建完成，可以用 Flume 采集工具将数据采集到 HDFS 当中。

11.4.5 日志信息说明

1. 数据产生

用户在浏览电商网站时，会触发很多事件，这时在前端页面设置数据埋点，会触发 Ajax 异步请求，将数据传送到后台。

2. 数据样例

数据样例如图 11.6 所示。

图 11.6　数据样例

3. 数据描述

（1）文件头，共 8 字节。

① 第一个起始 IP 的绝对偏移量，4 字节。

② 最后一个起始 IP 的绝对偏移量，4 字节。

（2）结束地址/国家/地区记录区，4 字节。IP 地址后跟的每一条记录分成两个部分：国家记录、地区记录。

（3）起始地址/结束地址偏移记录区。每条记录 7 字节，按照起始地址从小到大排列。

① 起始 IP 地址，4 字节。

② 结束 IP 地址的绝对偏移量，3 字节。注意：这个文件里的 IP 地址和所有的偏移量均采用小端字节序，而 Java 采用大端字节序，需要转换。

11.5 数据清洗

11.5.1 分析需要清洗的数据

通过 Flume 采集到 HDFS 中的数据为原生数据，也称为不规整数据，即目前来说，该数据的格式还无法满足我们对数据处理的基本要求，需要对其进行预处理，将其转化为我们所需要的较为规整的数据，以方便后面的统计分析。

11.5.2 解析数据格式转换

采集到的原始数据需要转换格式，包括 IP 地址解析、浏览器信息解析、服务器时间解析等。

（1）将 IP 地址解析为国家、省、市。

（2）对浏览器信息进行清洗转换。

（3）对时间进行清洗。

格式转换后数据样例如图 11.7 所示。

```
中国 河南 郑州 1537166659604 e_e event%E7%9A%84category%E5%90%8D%E7%A7%B0 event%E7%9A%84action%E5%90%8D%E7%A7%B0 value1 value2 1245 1 websit
5A049A21-0F9B-47C0-8490-2A8C784A99D4&c_time=1537166703623&l=zh-CN& chrom 32.0 windows window 7 1600*900
192.168.216.1^A1537166660.076^A192.168.216.111^A/index.html?en=e_e&ca=event%E7%9A%84category%E5%90%8D%E7%A7%B0&ac=event%E7%9A%84act
ion%E5%90%8D%E7%A7%B0&ver=1&pl=website&sdk=js&u_ud=1DF5A695-19BD-4DEF-9F8B-597B7BB1B0B7&u_sd=5A049A21-0F9B-47C0-8490-2A8C784A99D4&c
_time=1537166704095&l=zh-CN&b_iev=Mozilla%2F5.0%20(Windows%20NT%206.1%3B%20WOW64)%20AppleWebKit%2F537.36%20(KHTML%2C%20like%20Gecko
)%20Chrome%2F58.0.3029.110%20Safari%2F537.36%20SE%202.X%20MetaSr%201.0&b_rst=1600*900
```

图 11.7　格式转换后数据样例

11.5.3 利用 MapReduce 清洗数据

清洗数据需要编写 MapReduce 程序，而 MapReduce 程序的编写又分为写 Mapper、Reducer、Job 三个基本的过程。在本项目中，要达到数据清洗的目只需要 Mapper，其输出的数据并不需要进一步由 Redcuer 汇总处理。因篇幅所限，本书仅对核心代码进行展示。

（1）首先编写 LogWritable 类封装 JavaBean 组件，把所有的数据字段全都放到这个类里面，相当于自定义一个数据类型，目的是将数据以指定的格式输出出去，以便后续程序调用。

```java
public class LogWritable implements Writable {
    private String ver;
    private String s_time;
    private String en;
    private String u_ud;
    private String u_mid;
    private String u_sd;
    private String c_time;
```

```java
        private String l;
        private String b_iev;
        private String b_rst;
        private String p_url;
        private String p_ref;
        private String tt;
        private String pl;
        private String ip;
        private String oid;
        private String on;
        private String cua;
        private String cut;
        private String pt;
        private String ca;
        private String ac;
        private String kv_;
        private String du;
        private String browserName;
        private String browserVersion;
        private String osName;
        private String osVersion;
        private String country;
        private String province;
        private String city;
        @Override
        public void write(DataOutput dataOutput) throws IOException {
            dataOutput.writeUTF(ver);
            dataOutput.writeUTF(s_time);
            dataOutput.writeUTF(en);
            dataOutput.writeUTF(u_ud);
            dataOutput.writeUTF(u_mid);
            dataOutput.writeUTF(u_sd);
            dataOutput.writeUTF(c_time);
            dataOutput.writeUTF(l);
            dataOutput.writeUTF(b_iev);
            dataOutput.writeUTF(b_rst);
            dataOutput.writeUTF(p_url);
            dataOutput.writeUTF(p_ref);
            dataOutput.writeUTF(tt);
            dataOutput.writeUTF(pl);
            dataOutput.writeUTF(ip);
            dataOutput.writeUTF(oid);
            dataOutput.writeUTF(on);
            dataOutput.writeUTF(cua);
            dataOutput.writeUTF(cut);
            dataOutput.writeUTF(pt);
            dataOutput.writeUTF(ca);
            dataOutput.writeUTF(ac);
            dataOutput.writeUTF(kv_);
            dataOutput.writeUTF(du);
            dataOutput.writeUTF(browserName);
            dataOutput.writeUTF(browserVersion);
            dataOutput.writeUTF(osName);
            dataOutput.writeUTF(osVersion);
            dataOutput.writeUTF(country);
            dataOutput.writeUTF(province);
```

```java
            dataOutput.writeUTF(city);
        }
        @Override
        public void readFields(DataInput dataInput) throws IOException {
            this.ver= dataInput.readUTF();
            this.s_time= dataInput.readUTF();
            this.en= dataInput.readUTF();
            this.u_ud= dataInput.readUTF();
            this.u_mid= dataInput.readUTF();
            this.u_sd= dataInput.readUTF();
            this.c_time= dataInput.readUTF();
            this.l= dataInput.readUTF();
            this.b_iev= dataInput.readUTF();
            this.b_rst= dataInput.readUTF();
            this.p_url= dataInput.readUTF();
            this.p_ref= dataInput.readUTF();
            this.tt= dataInput.readUTF();
            this.pl= dataInput.readUTF();
            this.ip= dataInput.readUTF();
            this.oid= dataInput.readUTF();
            this.on= dataInput.readUTF();
            this.cua= dataInput.readUTF();
            this.cut= dataInput.readUTF();
            this.pt= dataInput.readUTF();
            this.ca= dataInput.readUTF();
            this.ac= dataInput.readUTF();
            this.kv_= dataInput.readUTF();
            this.du= dataInput.readUTF();
            this.browserName= dataInput.readUTF();
            this.browserVersion= dataInput.readUTF();
            this.osName= dataInput.readUTF();
            this.osVersion= dataInput.readUTF();
            this.country= dataInput.readUTF();
            this.province= dataInput.readUTF();
            this.city= dataInput.readUTF();
        }
    //这里需要对所有变量生成Get()、Set()方法，略
        @Override
        public String toString() {
            return ver + "\u0001"+s_time+"\001"+en+"\001"+u_ud+"\001"+
                    u_mid+"\001"+u_sd+"\001"+c_time+"\001"+l+"\001"+
                    b_iev+"\001"+b_rst+"\001"+p_url+"\001"+p_ref+"\001"+
                    tt+"\001"+pl+"\001"+ip+"\001"+oid+"\001"+on+"\001"+
                    cua+"\001"+cut+"\001"+pt+"\001"+ca+"\001"+ac+"\001"+
                    kv_+"\001"+du+"\001"+browserName+"\001"+browserVersion+"\001"+
            osName+"\001"+osVersion+"\001"+country+"\001"+province+"\001"+city;    }
}
```

通过对LogWritable类的处理，可以将数据调整为项目需要的格式进行输出。

（2）编写枚举类。

```java
/**
 * @Description 整个项目日志的常量类
 **/
public class Constants {
    * 事件的枚举
```

```java
public enum EventEnum{
    LANUCH(1,"lanuch event","e_l"),
    PAGEVIEW(2,"pageview event","e_pv"),
    EVENT(3,"event name","e_e"),
    CHARGEREQUEST(4,"charge request event","e_crt"),
    CHARGESUCCESS(5,"charge success","e_cs"),
    CHARGEREFUND(6,"charge refund","e_cr")
    ;
    public int id;
    public String name;
    public String alias;
    EventEnum(int id, String name, String alias) {
        this.id = id;
        this.name = name;
        this.alias = alias;
    }
    //根据名称获取枚举
    public static EventEnum valueOfAlias(String alia){
        for (EventEnum event : values()){
            if(event.alias.equals(alia)){
                return event;
            }
        }
        throw new RuntimeException("该alias没有对应的枚举.alias:"+alia);
    }
}
public static final String LOG_VERSION = "ver";
public static final String LOG_SERVER_TIME = "s_time";
public static final String LOG_EVENT_NAME = "en";
public static final String LOG_UUID = "u_ud";
public static final String LOG_MEMBER_ID = "u_mid";
public static final String LOG_SESSION_ID = "u_sd";
public static final String LOG_CLIENT_TIME = "c_time";
public static final String LOG_LANGUAGE = "l";
public static final String LOG_USERAGENT = "b_iev";
public static final String LOG_RESOLUTION = "b_rst";
public static final String LOG_CURRENT_URL = "p_url";
public static final String LOG_PREFFER_URL = "p_ref";
public static final String LOG_TITLE = "tt";
public static final String LOG_PLATFORM = "pl";
public static final String LOG_IP = "ip";
 * 和订单相关
public static final String LOG_ORDER_ID = "oid";
public static final String LOG_ORDER_NAME = "on";

public static final String LOG_CURRENCY_AMOUTN = "cua";
public static final String LOG_CURRENCY_TYPE = "cut";
public static final String LOG_PAYMENT_TYPE = "pt";
/**
 * 事件相关
 * 点击：点击事件 category 转发
 * 下单：下单事件
 */
public static final String LOG_EVENT_CATEGORY = "ca";
```

```java
    public static final String LOG_EVENT_ACTION = "ac";
    public static final String LOG_EVENT_KV = "kv_";
    public static final String LOG_EVENT_DURATION = "du";
    /**
     * 浏览器相关
     */
    public static final String LOG_BROWSER_NAME = "browserName";
    public static final String LOG_BROWSER_VERSION = "browserVersion";
    public static final String LOG_OS_NAME = "osName";
    public static final String LOG_OS_VERSION = "osVersion";
    /**
     * 地域相关
     */
    public static final String LOG_COUNTRY = "country";
    public static final String LOG_PROVINCE = "province";
    public static final String LOG_CITY = "city";
}
```

通过编写枚举类,给变量重新命名,便于项目代码的编写。

(3)编写 Mapper 程序代码。

```java
public class EtlToHdfsMapper extends
Mapper<LongWritable,Text,LogWritable,NullWritable> {
    private static final Logger logger = Logger.getLogger(EtlToHdfsMapper.class);
    private LogWritable k = new LogWritable();
    private int inputRecords,filterRecords,outputRecords = 0;
    @Override
    protected void map(LongWritable key, Text value, Context context) throws IOException,
InterruptedException {
        try {
            String line = value.toString();
            this.inputRecords ++;
            //空行处理
            if(StringUtils.isEmpty(line)){
                this.filterRecords ++;
                return ;
            }
            //调用 LogUtil 中的 parseLog()方法,返回 Map,然后循环 Map 将数据分别输出
            //也可以根据事件来分别输出
            Map<String,String> map = LogUtil.parserLog(line);
            //获取事件名
            String eventName = map.get(Constants.LOG_EVENT_NAME);
            Constants.EventEnum event = Constants.EventEnum.valueOfAlias(eventName);
            switch (event){
                case LANUCH:
                case EVENT:
                case PAGEVIEW:
                case CHARGEREQUEST:
                case CHARGESUCCESS:
                case CHARGEREFUND:
                    handleLog(map,context); //处理输出
                    break;
                default:
                    break;
            }
```

```java
            } catch (Exception e) {
                this.filterRecords ++;
                logger.warn("处理Mapper写出数据的时候异常.",e);
            }
        }
        /**
         *将Map中的所有k-v对进行输出
         * @param map
         * @param context
         */
        private void handleLog(Map<String,String> map, Context context) {

            try {
                //Map循环
                for(Map.Entry<String,String> en : map.entrySet()){
                    this.k.setB_iev(en.getValue());
                    switch (en.getKey()){
                        case "ver": this.k.setIp(en.getValue()); break;
                        case "s_time": this.k.setS_time(en.getValue()); break;
                        case "en": this.k.setEn(en.getValue()); break;
                        case "u_ud": this.k.setU_ud(en.getValue()); break;
                        case "u_mid": this.k.setU_mid(en.getValue()); break;
                        case "u_sd": this.k.setU_sd(en.getValue()); break;
                        case "c_time": this.k.setC_time(en.getValue()); break;
                        case "l": this.k.setL(en.getValue()); break;
                        case "b_iev": this.k.setB_iev(en.getValue()); break;
                        case "b_rst": this.k.setB_rst(en.getValue()); break;
                        case "p_url": this.k.setP_url(en.getValue()); break;
                        case "p_ref": this.k.setP_ref(en.getValue()); break;
                        case "tt": this.k.setTt(en.getValue()); break;
                        case "pl": this.k.setPl(en.getValue()); break;
                        case "ip": this.k.setIp(en.getValue()); break;
                        case "oid": this.k.setOid(en.getValue()); break;
                        case "on": this.k.setOn(en.getValue()); break;
                        case "cua": this.k.setCua(en.getValue()); break;
                        case "cut": this.k.setCut(en.getValue()); break;
                        case "pt": this.k.setPt(en.getValue()); break;
                        case "ca": this.k.setCa(en.getValue()); break;
                        case "ac": this.k.setAc(en.getValue()); break;
                        case "kv_": this.k.setKv_(en.getValue()); break;
                        case "du": this.k.setDu(en.getValue()); break;
                        case "browserName": this.k.setBrowserName(en.getValue()); break;
                        case "browserVersion": this.k.setBrowserVersion(en.getValue()); break;
                        case "osName": this.k.setOsName(en.getValue()); break;
                        case "osVersion": this.k.setOsVersion(en.getValue()); break;
                        case "country": this.k.setCountry(en.getValue()); break;
                        case "province": this.k.setProvince(en.getValue()); break;
                        case "city": this.k.setCity(en.getValue()); break;
                    }
                }
                this.outputRecords ++;
                context.write(k,NullWritable.get());
            } catch (Exception e) {
                logger.warn("etl最终输出异常",e);
            }
```

```
    }
    @Override
    protected void cleanup(Context context) throws IOException, InterruptedException {
        logger.info("inputRecords:"+this.inputRecords+"    filterRecords:"+filterRecords+" outputRecords:"+outputRecords);
    }
}
```

采集到的原生数据通过 Map 端处理，转换为后续模块开发中所需要的字段信息。

（4）编写 Job 类。编写 Job 驱动类，提交 MapReduce 程序，设置数据目录。

原数据存储目录：/log/09/18。

原数据存储目录：/log/09/19。

清洗后的存储目录：/ods/09/18。

清洗后的存储目录：/ods/09/19。

```
public class EtlToHdfsRunner implements Tool {
    private static final Logger logger = Logger.getLogger(EtlToHdfsRunner.class);
    private Configuration conf = new Configuration();
    //主函数
    public static void main(String[] args) {
        try {
            ToolRunner.run(new Configuration(),new EtlToHdfsRunner(),args);
        } catch (Exception e) {
            logger.warn("执行etl to hdfs 异常.",e);
        }
    }
    @Override
    public void setConf(Configuration conf) {
        conf = this.conf;
    }
    @Override
    public Configuration getConf() {
        return this.conf;
    }
    @Override
    public int run(String[] args) throws Exception {
        Configuration conf = getConf();
        //获取-d之后的日期并存储到 conf 中，如果没有-d 或者日期不合法则以昨天为默认值
        this.handleArgs(conf,args);
        //获取 Job
        Job job= Job.getInstance(conf,"etl to hdfs");
        job.setJarByClass(EtlToHdfsRunner.class);
        //设置 Map 相关
        job.setMapperClass(EtlToHdfsMapper.class);
        job.setMapOutputKeyClass(LogWritable.class);
        job.setMapOutputValueClass(NullWritable.class);
        //没有 Reduce
        job.setNumReduceTasks(0);
        //设置输入输出
        this.handleInputOutpu(job);
        return job.waitForCompletion(true)?1:0;
    }
    /**
```

```java
     *
     * @param configuration
     * @param args
     */
    private void handleArgs(Configuration configuration, String[] args) {
        String date = null;
        if(args.length > 0){
            //循环args
            for(int i = 0 ; i<args.length;i++){
                //判断参数中是否有-d
                if(args[i].equals("-d")){
                    if(i+1 <= args.length){
                        date = args[i+1];
                        break;
                    }
                }
            }
            //判断
            if(StringUtils.isEmpty(date)){
                date = TimeUtil.getYesterday();
            }
            //将date存储到conf中
            conf.set(GlobalConstants.RUNNING_DATE,date);
        }
    }
    /**
     * 设置输入输出
     * @param job
     */
    private void handleInputOutpu(Job job) {
        String [] fields = job.getConfiguration().get(GlobalConstants.RUNNING_DATE).split("-");
        String month = fields[1];
        String day = fields[2];
        try {
            FileSystem fs = FileSystem.get(job.getConfiguration());
            Path inpath = new Path("/log/"+month+"/"+day);
            Path outpath = new Path("/ods/"+month+"/"+day);
            if(fs.exists(inpath)){
                FileInputFormat.addInputPath(job,inpath);
            } else {
                throw  new RuntimeException("输入路径不存储在.inpath:"+inpath.toString());
            }
            //设置输出
            if(fs.exists(outpath)){
                fs.delete(outpath,true);
            }
            //设置输出
            FileOutputFormat.setOutputPath(job,outpath);
        } catch (IOException e) {
            logger.warn("设置输入输出路径异常.",e);
        }
```

        }
    }

（5）将写好的程序打成 jar 包提交到集群上运行，命令如下。

```
Yarn jar /home/Hadoop/phone_analystic-1.0.jar
com.phone.etl.mr.EtlToHdfsRunner -d 2018-9-18
```

11.6 使用数据仓库进行数据分析

使用 MapReduce 满足业务需求代码量过大，比较烦琐，所以开发人员一般情况下使用 Hive 数据仓库，写 SQL 语句实现功能。下面从创建数据库开始，带领大家一步步实现本项目中的各模块业务功能。

11.6.1 事件板块数据分析

（1）创建数据库。

```
create database log_phone;
use log_phone;
```

（2）编写事件的维度类和修改操作基础维度服务。

（3）编写 UDF 函数。

UDF 代码如下。

```java
package com.phone.analystic.hive;
import com.phone.analystic.modle.base.EventDimension;
import com.phone.analystic.mr.service.IDimension;
import com.phone.analystic.mr.service.impl.IDimensionImpl;
import com.phone.common.GlobalConstants;
import org.apache.commons.lang.StringUtils;
import org.apache.hadoop.hive.ql.exec.UDF;
/**
 * @ClassName EventDimensionUdf
 * @Author lyd
 * @Date $ $
 * @Vesion 1.0
 * @Description 获取事件维度的 id
 **/
public class EventDimensionUdf extends UDF {
    IDimension iDimension = new IDimensionImpl();
    /**
     * @param category
     * @param action
     * @return 事件维度的 id
     */
    public int evaluate(String category,String action){
private static Logger logger=Logger.getLogger(EventDimensionUdf.class);
        if(StringUtils.isEmpty(category)){
            category = action = GlobalConstants.DEFAULT_VALUE;
        }
        if(StringUtils.isEmpty(action)){
            action = GlobalConstants.DEFAULT_VALUE;
        }
```

```
        int id = -1;
        try {
            EventDimension ed = new EventDimension(category,action);
            id = iDimension.getDiemnsionIdByObject(ed);
        } catch (Exception e) {
logger.error(e.getMessage(),e);
        }
        return id;
    }
}
```

将编写好的 UDF 函数代码提交到 HDFS 集群中。

```
create function phone_event as
'com.phone.analystic.hive.EventDimensionUdf'
using jar 'hdfs://qianfeng01:9000/phone/udfjars/phone_analystic-1.0.jar';
create function phone_date as
'com.phone.analystic.hive.DateDimensionUdf'
using jar 'hdfs://qianfeng01:9000/phone/udfjars/phone_analystic-1.0.jar';
create function phone_platform as
'com.phone.analystic.hive.PlatformDimensionUdf'
using jar 'hdfs://qianfeng01:9000/phone/udfjars/phone_analystic-1.0.jar';
```

（4）创建元数据对应的临时表。

```
create external table if not exists phone_tmp(
ver string,
s_time string,
en string,
u_ud string,
u_mid string,
u_sd string,
c_time string,
l string,
b_iev string,
b_rst string,
p_url string,
p_ref string,
tt string,
pl string,
ip String,
oid String,
'on' String,
cua String,
cut String,
pt String,
ca String,
ac String,
kv_ String,
du String,
browserName String,
browserVersion String,
osName String,
osVersion String,
country String,
province String,
```

```
city string
)
partitioned by(month string,day string) ;
```

（5）导入数据。

```
load data inpath '/ods/09/19' into table phone_tmp
partition(month=09,day=19);
```

（6）创建 Hive 表。

```
create external table if not exists phone(
ver string,
s_time string,
en string,
u_ud string,
u_mid string,
u_sd string,
c_time string,
l string,
b_iev string,
b_rst string,
p_url string,
p_ref string,
tt string,
pl string,
ip String,
oid String,
'on' String,
cua String,
cut String,
pt String,
ca String,
ac String,
kv_ String,
du String,
browserName String,
browserVersion String,
osName String,
osVersion String,
country String,
province String,
city string
)
partitioned by(month string,day string)
stored as orc;
set hive.exec.local.mode=true;
from phone_tmp
insert into phone partition(month=09,day=19)
select
ver,
s_time,
en,
u_ud,
u_mid,
u_sd,
c_time,
```

```
    l,
    b_iev,
    b_rst,
    p_url,
    p_ref,
    tt,
    pl,
    ip,
    oid,
    'on',
    cua,
    cut,
    pt,
    ca,
    ac,
    kv_,
    du,
    browserName,
    browserVersion,
    osName,
    osVersion,
    country,
    province,
    city
    where month = 9
    and day = 19;
```

（7）在 Hive 中创建和 MySQL 最终结果表一样的临时表。

```
CREATE TABLE if not exists 'stats_event' (
  'platform_dimension_id' int,
  'date_dimension_id' int,
  'event_dimension_id' int,
  'times' int,
  'created' String
);
语句:
set hive.exec.mode.local.auto=true;
set hive.groupby.skewindata=true;
from(
select
from_unixtime(cast(p.s_time/1000 as bigint),"yyyy-MM-dd") as dt,
p.pl as pl,
p.ca as ca,
p.ac as ac,
count(*) as ct
from phone p
where p.month = 9
and p.day = 19
and en = 'e_l'
group by from_unixtime(cast(p.s_time/1000 as bigint),"yyyy-MM-dd"),p.pl,p.ca,p.ac
) as tmp
insert overwrite table stats_event
select phone_platform(pl),phone_date(dt),phone_event(ca,ac),ct,dt;
```

(8)扩展维度。

```
set hive.exec.mode.local.auto=true;
set hive.groupby.skewindata=true;
with tmp as(
select
from_unixtime(cast(l.s_time/1000 as bigint),"yyyy-MM-dd") as dt,
l.pl as pl,
l.ca as ca,
l.ac as ac
from phone l
where month = 9
and day = 19
and l.en = 'e_1'
and l.s_time <> 'null'
)
from (
    select dt as dt,pl as pl,ca as ca,ac as ac,count(1) as ct from tmp group by dt,pl,ca,ac
union all
    select dt as dt,pl as pl,ca as ca,'all' as ac,count(1) as ct from tmp group by dt,pl,ca
union all
    select dt as dt,'all' as pl,ca as ca,ac as ac,count(1) as ct from tmp group by dt,ca,ac
union all
    select dt as dt,'all' as pl,ca as ca,'all' as ac,count(1) as ct from tmp group by dt,ca
) as tmp1
insert overwrite table stats_event
select phone_date(dt),phone_platform(pl),phone_event(ca,ac),sum(ct),'2018-09-19'
group by pl,dt,ca,ac;
```

(9)输入测试数据。

```
3    1    1
3    1    3
3    3    1
3    3    3
```

(10)Sqoop 导出数据。

```
sqoop export --connect jdbc:mysql://qf01:3306/result \
--username root --password 911208 -m 1 \
--table stats_event --export-dir hdfs://cch-host:9000/user/hive/warehouse/log_phone.db/stats_event/* \
--input-fields-terminated-by "\\01" --update-mode allowinsert \
--update-key date_dimension_id,platform_dimension_id,event_dimension_id \
;
```

(11)结果展示。

```
1    3    订单事件|订单产生    1    2018-09-19
1    3    订单事件|订单支付    1    2018-09-19
1    3    点击|赞    200    2018-09-19
```

11.6.2 订单板块数据分析

(1)创建维度类,并创建对应的 UDF 方法。

UDF 代码如下。

```java
* @Description 获取平台维度的id
public class PlatformDimensionUdf extends UDF {
    IDimension iDimension = new IDimensionImpl();
     * @return 平台维度的id
    public int evaluate(String platform){
        if(StringUtils.isEmpty(platform)){
            platform = GlobalConstants.DEFAULT_VALUE;
        }
        int id = -1;
        try {
            PlatformDimension pl = new PlatformDimension(platform);
            id = iDimension.getDiemnsionIdByObject(pl);
        } catch (Exception e) {
          e.printStackTrace();
        }
        return id;
    }
    public static void main(String[] args) {
        System.out.println(new PlatformDimensionUdf().evaluate("website"));
    }
}
```

将编写好的UDF函数代码提交到HDFS集群中。

```
create function convert_pay as
 'com.qianfeng.analystic.hive.PaymentTypeDimensionUdf' using jar
 'hdfs://qf01:9000/lslda/udf/jars/LSLogAnalystic-1.0.jar';
create function convert_currency as
 'com.qianfeng.analystic.hive.CurrencyTypeDimensionUdf' using jar
 'hdfs://qf01:9000/lslda/udf/jars/LSLogAnalystic-1.0.jar';
```

（2）创建表。

```
create table if not exists stats_order_tmp(
    'date_dimension_id' int,
    'platform_dimension_id' int,
    'currency_type_dimension_id' int,
    'payment_type_dimension_id' int,
    'ct' int,
    'created' string
);
create table if not exists stats_order_tmp1(
    'date_dimension_id' int,
    'platform_dimension_id' int,
    'currency_type_dimension_id' int,
    'payment_type_dimension_id' int,
    'orders' int,
    'success_orders' int,
    'refund_orders' int,
    'created' string
);
create table if not exists stats_order_tmp2(
    'date_dimension_id' int,
    'platform_dimension_id' int,
    'currency_type_dimension_id' int,
    'payment_type_dimension_id' int,
    'orders_amount' double,
```

```
    'success_orders_amount' double,
    'refund_orders_amount' double,
    'created' string
);
```

（3）写指标。

总的订单数量。

```
from(
select
from_unixtime(cast(l.s_time/1000 as bigint),"yyyy-MM-dd") as dt,
l.pl as pl,
l.cut as cut,
l.pt as pt,
count(distinct l.o_id) as ct
from logs l
where l.month = 8
and day = 18
and l.o_id is not null
and l.o_id <> 'null'
and l.en = 'e_crt'
group by from_unixtime(cast(l.s_time/1000 as bigint),"yyyy-MM-dd"),pl,cut,pt
) as tmp
insert overwrite table stats_order_tmp
select    convert_date(dt),convert_platform(pl),convert_currency(cut),convert_pay(pt),sum(ct),dt
group by dt,pl,cut,pt
;
```

Sqoop 导出数据。

```
sqoop export --connect jdbc:mysql://qf01:3306/result \
--username root --password root \
--table stats_order --export-dir /hive/stats_order_tmp/* \
--input-fields-terminated-by "\\01" --update-mode allowinsert \
--update-key date_dimension_id,platform_dimension_id,currency_type_dimension_id,payment_type_dimension_id \
--columns 'date_dimension_id,platform_dimension_id,currency_type_dimension_id,payment_type_dimension_id,orders,created' \
;
```

成功支付的订单数量。

```
from(
select
from_unixtime(cast(l.s_time/1000 as bigint),"yyyy-MM-dd") as dt,
l.pl as pl,
l.cut as cut,
l.pt as pt,
count(distinct l.o_id) as ct
from logs l
where l.month = 8
and day = 18
and l.o_id is not null
and l.o_id <> 'null'
```

```
    and l.en = 'e_cs'
    group by from_unixtime(cast(l.s_time/1000 as bigint),"yyyy-MM-dd"),pl,cut,pt
) as tmp
insert overwrite table stats_order_tmp
select convert_date(dt),convert_platform(pl),convert_currency(cut),convert_pay(pt),sum(ct),dt
group by dt,pl,cut,pt
;
```

Sqoop 导出数据。

```
sqoop export --connect jdbc:mysql://qf01:3306/result \
--username root --password root \
--table stats_order --export-dir /hive/stats_order_tmp/* \
--input-fields-terminated-by "\\01" --update-mode allowinsert \
--update-key date_dimension_id,platform_dimension_id,currency_type_dimension_id,payment_type_dimension_id \
--columns 'date_dimension_id,platform_dimension_id,currency_type_dimension_id,payment_type_dimension_id,success_orders,created' \
;
```

退款成功的订单数量。

```
from(
select
from_unixtime(cast(l.s_time/1000 as bigint),"yyyy-MM-dd") as dt,
l.pl as pl,
l.cut as cut,
l.pt as pt,
count(distinct l.o_id) as ct
from logs l
where l.month = 8
and day = 18
and l.o_id is not null
and l.o_id <> 'null'
and l.en = 'e_cr'
group by from_unixtime(cast(l.s_time/1000 as bigint),"yyyy-MM-dd"),pl,cut,pt
) as tmp
insert overwrite table stats_order_tmp
select convert_date(dt),convert_platform(pl),convert_currency(cut),convert_pay(pt),sum(ct),dt
group by dt,pl,cut,pt
;
```

Sqoop 导出数据。

```
sqoop export --connect jdbc:mysql://qf01:3306/result \
--username root --password root -m 1 \
--table stats_order --export-dir /hive/stats_order_tmp/* \
--input-fields-terminated-by \\01
 --update-mode allowinsert \
--update-key date_dimension_id,platform_dimension_id,currency_type_dimension_id,payment_type_dimension_id \
--columns 'date_dimension_id,platform_dimension_id,currency_type_dimension_id,payment_type_dimension_id,refund_orders,created' \
;
```

计算订单数量写成一条语句如下。

```sql
with tmp as(
select
from_unixtime(cast(l.s_time/1000 as bigint),"yyyy-MM-dd") as dt,
l.pl as pl,
l.cut as cut,
l.pt as pt,
l.en as en,
if((case when l.en = 'e_crt' then count(distinct l.o_id) end) is null,0,(case when l.en = 'e_crt' then count(distinct l.o_id) end))as orders,
if((case when l.en = 'e_cs' then count(distinct l.o_id) end) is null,0,(case when l.en = 'e_cs' then count(distinct l.o_id) end))as success_orders,
if((case when l.en = 'e_cr' then count(distinct l.o_id) end) is null,0,(case when l.en = 'e_cr' then count(distinct l.o_id) end))as refund_orders
from logs l
where l.month = 8
and day = 18
and l.o_id is not null
and l.o_id <> 'null'
group by from_unixtime(cast(l.s_time/1000 as bigint),"yyyy-MM-dd"),pl,cut,pt,l.en
)
from(
select dt as dt1,pl as pl ,cut as cut,pt as pt,orders as orders,0 as success_orders,0 as refund_orders,dt from tmp where en = 'e_crt'
union all
select dt as dt1,pl as pl ,cut as cut,pt as pt,0 as orders,success_orders as success_orders,0 as refund_orders,dt from tmp where en = 'e_cs'
union all
select dt as dt1,pl as pl ,cut as cut,pt as pt,0 as orders,0 as success_orders,refund_orders as refund_orders,dt from tmp where en = 'e_cr'
) as tmp1
insert overwrite table stats_order_tmp1
select  convert_date(dt1),convert_platform(pl),convert_currency(cut),convert_pay(pt),sum(orders),sum(success_orders),sum(refund_orders),dt1
group by dt1,pl,cut,pt
;
```

计算三种金额写成一条语句如下。

```sql
from(
select
from_unixtime(cast(l.s_time/1000 as bigint),"yyyy-MM-dd") as dt,
l.pl as pl,
l.cut as cut,
l.pt as pt,
if((case when l.en = 'e_crt' then sum(l.cua) end) is null,0,(case when l.en = 'e_crt' then sum(l.cua) end))as orders_amount,
if((case when s.en = 'e_cs' then sum(l.cua) end) is null,0,(case when s.en = 'e_cs' then sum(l.cua) end))as success_orders_amount,
if((case when r.en = 'e_cr' then sum(l.cua) end) is null,0,(case when r.en = 'e_cr' then sum(l.cua) end))as refund_orders_amount
from logs l
left join logs s
on s.o_id = l.o_id and s.en = 'e_cs'
left join logs r
on r.o_id = s.o_id and r.en = 'e_cr'
where l.month = 8
and l.day = 18
```

```
        and l.o_id is not null
        and l.o_id <> 'null'
        and l.en = 'e_crt'
    group by from_unixtime(cast(l.s_time/1000 as bigint),"yyyy-MM-dd"),l.pl,l.cut,l.pt,l.en,
s.en,r.en
    ) as tmp
    insert overwrite table stats_order_tmp2
    select convert_date(dt),convert_platform(pl),convert_currency(cut),convert_pay(pt),sum
(orders_amount),sum(success_orders_amount),sum(refund_orders_amount),dt
    group by dt,pl,cut,pt
;
```

Sqoop 导出数据。

```
sqoop export --connect jdbc:mysql://qf01:3306/result \
--username root --password root -m 1 \
--table stats_order --export-dir /hive/stats_order_tmp2/* \
--input-fields-terminated-by "\\01" --update-mode allowinsert \
--update-key
date_dimension_id,platform_dimension_id,currency_type_dimension_id,payment_type_dimension
_id \
    --columns 'date_dimension_id,platform_dimension_id,currency_type_dimension_id,payment_
type_dimension_id,order_amount,revenue_amount,refund_amount,created' \
;
```

11.6.3 时间板块数据分析

（1）数据抽取（抽取该模块中需要的字段即可）。

```
load data inpath '/ods/month=08/day=18' into table logs
partition(month=08,day=18);
```

（2）创建最终结果表。

```
CREATE TABLE IF NOT EXISTS 'stats_view_depth' (
  'platform_dimension_id' int,
  'data_dimension_id' int,
  'kpi_dimension_id' int,
  'pv1' int,
  'pv2' int,
  'pv3' int,
  'pv4' int,
  'pv5_10' int,
  'pv10_30' int,
  'pv30_60' int,
  'pv60pluss' int,
  'created' string
);
```

（3）创建临时表。

```
CREATE TABLE IF NOT EXISTS 'stats_view_depth_tmp' (
dt string,
pl string,
col string,
ct int
```

```
);
2018-09-19  website  pv1  10
2018-09-19  website  pv2  200
3    1    2    10   200  0    0    0    0    2018-09-19
3    1    2    10   0    0    0    0    0    2018-09-19
3    1    2    0    200  0    0    0    0    2018-09-19
3    1    2    10   0    0    0    0    0    2018-09-19
```

(4) SQL 语句。

```
set hive.exec.mode.local.auto=true;
set hive.groupby.skewindata=true;
from(
select
from_unixtime(cast(l.s_time/1000 as bigint),"yyyy-MM-dd") as dt,
l.pl as pl,
l.u_ud as uid,
(case
when count(l.p_url) = 1 then "pv1"
when count(l.p_url) = 2 then "pv2"
when count(l.p_url) = 3 then "pv3"
when count(l.p_url) = 4 then "pv4"
when count(l.p_url) < 10 then "pv5_10"
when count(l.p_url) < 30 then "pv10_30"
when count(l.p_url) < 60 then "pv30_60"
else "pv60pluss"
end) as pv
from phone l
where month = 09
and day = 19
and l.p_url <> 'null'
and l.pl is not null
group by from_unixtime(cast(l.s_time/1000 as bigint),"yyyy-MM-dd"),pl,u_ud
) as tmp
insert overwrite table stats_view_depth_tmp
select dt,pl,pv,count(distinct uid) as ct
where uid is not null
group by dt,pl,pv
;
set hive.exec.mode.local.auto=true;
set hive.groupby.skewindata=true;
with tmp as(
    select dt,pl as pl,ct as pv1,0 as pv2,0 as pv3,0 as pv4,0 as pv5_10,0 as pv10_30,0 as pv30_60,0 as pv60pluss from stats_view_depth_tmp where col = 'pv1' union all
    select dt,pl as pl,0 as pv1,ct as pv2,0 as pv3,0 as pv4,0 as pv5_10,0 as pv10_30,0 as pv30_60,0 as pv60pluss from stats_view_depth_tmp where col = 'pv2' union all
    select dt,pl as pl,0 as pv1,0 as pv2,ct as pv3,0 as pv4,0 as pv5_10,0 as pv10_30,0 as pv30_60,0 as pv60pluss from stats_view_depth_tmp where col = 'pv3' union all
    select dt,pl as pl,0 as pv1,0 as pv2,0 as pv3,ct as pv4,0 as pv5_10,0 as pv10_30,0 as pv30_60,0 as pv60pluss from stats_view_depth_tmp where col = 'pv4' union all
    select dt,pl as pl,0 as pv1,0 as pv2,0 as pv3,0 as pv4,ct as pv5_10,0 as pv10_30,0 as pv30_60,0 as pv60pluss from stats_view_depth_tmp where col = 'pv5_10' union all
    select dt,pl as pl,0 as pv1,0 as pv2,0 as pv3,0 as pv4,0 as pv5_10,ct as pv10_30,0 as pv30_60,0 as pv60pluss from stats_view_depth_tmp where col = 'pv10_30' union all
    select dt,pl as pl,0 as pv1,0 as pv2,0 as pv3,0 as pv4,0 as pv5_10,0 as pv10_30,ct as pv30_60,0 as pv60pluss from stats_view_depth_tmp where col = 'pv30_60' union all
```

```
      select dt,pl as pl,0 as pv1,0 as pv2,0 as pv3,0 as pv4,0 as pv5_10,0 as pv10_30,0 as pv30_60,ct
as pv60pluss from stats_view_depth_tmp where col = 'pv60pluss'
    )
    from tmp
    insert overwrite table stats_view_depth
    select phone_date(dt),phone_platform(pl),2,sum(pv1),sum(pv2),sum(pv3),sum(pv4),sum (pv5_
10),sum(pv10_30),sum(pv30_60),sum(pv60pluss),dt
    group by dt,pl
    ;
```

（5）Sqoop 导出数据。

```
sqoop export --connect jdbc:mysql://qf01:3306/result \
 --username root --password root \
 --table stats_view_depth --export-dir /hive/log_phone.db/stats_view_depth/* \
 --input-fields-terminated-by "\\01" --update-mode allowinsert \
 --update-key date_dimension_id,platform_dimension_id,kpi_dimension_id \
 ;
```

（6）用户角度下的访问深度。

```
2018-08-17 website http://localhost:8080/index.html 123
2018-08-17 website http://localhost:8080/index.html 123
2018-08-17 website http://localhost:8080/index.html 123
2018-08-17 website http://localhost:8080/index1.html 345
2018-08-17 website http://localhost:8080/index.html 345
```

（7）统计 PV 的值。

```
2018-08-18    website   2A6FB951-F4FC-4886-87C0-E9C9D47D2C5C    pv1
2018-08-18    website   D4289356-5BC9-47C4-8F7D-F16022833E7E    pv1
2018-08-18    website   pv1  2
2018-08-18    website   pv10_30  90
group by
2018-08-17 website 1 pv1
2018-08-17 website 1 pv1
2018-08-17 website 3 pv3
```

（8）将统计的 PV 的值和对应的 pv1、pv2 等存储到临时表。

```
dt string,
pl string,
col string,
value ''
```

11.7 可视化

11.7.1 ECharts 简介

ECharts（Enterprise Charts，商业级数据图表）是百度的一个开源的数据可视化工具，在业界获得了很多赞誉。下面介绍如何使用百度的开源框架 ECharts 完成数据可视化。ECharts 官方网站如图 11.8 所示。

图 11.8 ECharts 官方网站

11.7.2 ECharts 的优点

ECharts 官方网站提供了源码和说明文档，以及可供使用的大量图表，如图 11.9 所示。ECharts 的优点如下。

图 11.9 ECharts 提供大量图表

（1）开源软件，完全免费，采用漂亮的图形化界面。
（2）操作简单，内部封装了重要的 JS，可直接引用。
（3）提供了各种图表。
（4）兼容性好，基于 HTML5 渲染动画。

11.7.3 操作流程

下面以地图为例，对 ECharts 的操作流程进行简单阐述。
（1）向 HTML 中引入 JS 文件。

```
<script type="text/javascript" src="echarts-all.js"></script>
```

（2）在使用 ECharts 制图之前，首先要定义一个区域。

```
<body>
    <!-- 为 ECharts 准备一个具备大小（宽高）的 DOM -->
    <!-- 地图 map-->
    <div id="echarts_map" style="width: 600px;height:400px;"></div>
</body>
```

（3）写入随机数来测试数据。

```
function randomData() {
    return Math.round(Math.random()*500);
}
```

写入数据。

```
var mydata = [
    {name: '北京',value: '100' },{name: '天津',value: randomData() },
    {name: '上海',value: randomData() },{name: '重庆',value: randomData() },
    {name: '河北',value: randomData() },{name: '河南',value: randomData() },
    {name: '云南',value: randomData() },{name: '辽宁',value: randomData() },
    {name: '黑龙江',value: randomData() },{name: '湖南',value: randomData() },
    {name: '安徽',value: randomData() },{name: '山东',value: randomData() },
    {name: '新疆',value: randomData() },{name: '江苏',value: randomData() },
    {name: '浙江',value: randomData() },{name: '江西',value: randomData() },
    {name: '湖北',value: randomData() },{name: '广西',value: randomData() },
    {name: '甘肃',value: randomData() },{name: '山西',value: randomData() },
    {name: '内蒙古',value: randomData() },{name: '陕西',value: randomData() },
    {name: '吉林',value: randomData() },{name: '福建',value: randomData() },
    {name: '贵州',value: randomData() },{name: '广东',value: randomData() },
    {name: '青海',value: randomData() },{name: '西藏',value: randomData() },
    {name: '四川',value: randomData() },{name: '宁夏',value: randomData() },
    {name: '海南',value: randomData() },{name: '台湾',value: randomData() },
    {name: '香港',value: randomData() },{name: '澳门',value: randomData() }
];
```

配置属性，置入数据。

```
var optionMap = {
            backgroundColor: '#FFFFFF',
            title: {
                text: '全国地图大数据',
                subtext: '',
                x:'center'
            },
            tooltip : {
                trigger: 'item'
            },
            //左侧小导航图标
            visualMap: {
                show : true,
                x: 'left',
```

```
                y: 'center',
                splitList: [
                    {start: 500, end:600},{start: 400, end: 500},
                    {start: 300, end: 400},{start: 200, end: 300},
                    {start: 100, end: 200},{start: 0, end: 100},
                ],
                color: ['#5475f5',   '#9feaa5',   '#85daef','#74e2ca',   '#e6ac53',
'#9fb5ea']
            },
            //配置属性
            series: [{
                name: '数据',
                type: 'map',
                mapType: 'china',
                roam: true,
                label: {
                    normal: {
                        show: true   //省份名称
                    },
                    emphasis: {
                        show: false
                    }
                },
                data:mydata    //数据
            }]
        };
//初始化ECharts实例
var myChart = echarts.init(document.getElementById('echarts_map'));
//使用指定的配置项和数据显示图表
myChart.setOption(optionMap);
```

11.8 本章小结

问题是时代的声音，回答并指导解决问题是理论的根本任务。本章首先从开发背景、需求分析、开发环境、系统预览等层面对项目案例进行概括性介绍，然后分别进行数据采集、数据预处理、数据仓库、数据分析、数据导出各模块开发。通过本章的学习，应掌握整个项目的架构、流程，以及各模块在项目开发中的作用。

11.9 习题

思考题

（1）简述本章项目案例的开发背景以及系统架构。
（2）简述本章项目案例中数据仓库开发的实现思路。
（3）简述本章项目案例中数据分析板块的实现思路。

附录

本书 Hadoop 环境配置需要安装包,请扫描下方二维码关注"千问千知"微信公众号,在对话框输入"Hadoop 开发环境"即可领取安装包及安装步骤。

千问千知